Food Science and Technology

nova science publishers

www.novapublishers.com

Food Science and Technology

Nutraceuticals: Food Applications and Health Benefits
Anita Kumari, PhD
Gulab Singh, PhD (Editors)
2022. ISBN: 978-1-68507-488-3 (Hardcover)
2022. ISBN: 978-1-68507-512-5 (eBook)

A Closer Look at Polyphenolics
Peter Bertollini (Editor)
2022. ISBN: 978-1-68507-480-7 (eBook)
2021. ISBN: 978-1-68507-434-0 (Hardcover)

Food Processing: Advances in Research and Applications
Myriam Huijs (Editor)
2022. ISBN: 978-1-68507-570-5 (Hardcover)
2022. ISBN: 978-1-68507-581-1 (eBook)

Cheeses around the World:
Types, Production, Properties and Cultural and Nutritional Relevance
Ana Cristina Ferrão, Msc, Paula Maria dos Reis Correia,
Raquel de Pinho Ferreira Guiné, PhD (Editors)
2019. ISBN: 978-1-53615-418-4 (Hardcover)
2019. ISBN: 978-1-53615-419-1 (eBook)

Flour: Production, Varieties and Nutrition
María Dolores Torres Pérez (Editor)
2018. ISBN: 978-1-53613-761-3 (Hardcover)
2018. ISBN: 978-1-53613-762-0 (eBook)

More information about this series can be found at
https://novapublishers.com/product-category/series/food-science-and-technology-series/

Nirmala Sehrawat,
Diwakar Aggarwal,
Sushil Kumar Upadhyay
and Mukesh Yadav
Editors

Flavonoids

Dietary Sources, Biological Properties and Therapeutic Potential

nova
science publishers
www.novapublishers.com

DOI: https://doi.org/10.52305/TYXJ2027

Library of Congress Cataloging-in-Publication Data

Names: Sehrawat, Nirmala, editor.
Title: Flavonoids : dietary sources, biological properties and therapeutic potential / Nirmala Sehrawat, PHD (EDITOR), Assistant Professor, Department of Biotechnology, Maharishi, Markandeshwar (Deemed to be University), Mullana-Ambala, India, Diwakar Aggarwal, PHD (EDITOR), Assistant Professor, Department of Biotechnology, Maharishi Markandeshwar (Deemed to be University), Mullana-Ambala, India, Sushil Kumar Upadhyay, PHD (EDITOR), Assistant Professor, Department of Biotechnology, Maharishi Markandeshwar (Deemed to be University), Mullana-Ambala, India, Mukesh Yadav, PHD (EDITOR), Assistant Professor, Department of Biotechnology, Maharishi Markandeshwar (Deemed to be University), Mullana-Ambala, India.
Description: New York : Nova Science Publishers, [2022] | Series: Food science and technology | Includes bibliographical references and index.
Identifiers: LCCN 2022027593 (print) | LCCN 2022027594 (ebook) | ISBN 9781685079536 (hardcover) | ISBN 9798886970319 (adobe pdf)
Subjects: LCSH: Flavonoids.
Classification: LCC QP671.F52 F54 2022 (print) | LCC QP671.F52 (ebook) | DDC 615.3/21--dc23/eng/20220713
LC record available at https://lccn.loc.gov/2022027593
LC ebook record available at https://lccn.loc.gov/2022027594

Published by Nova Science Publishers, Inc. † New York

Contents

Preface

Dietary flavonoids have emerged as a safe and effective way of treating, controlling or managing various diseases. Various flavonoids have been investigated as natural, cost effective, alternate and combinatorial therapy for managing various diseases and have been found promising. Various functional foods have proven effective and beneficial for human heath due to the presence of flavonoids. Flavonoids have antimicrobial, anti-inflammatory and antioxidant properties. The antioxidant and anti-inflammatory properties result in various health benefits to consumers. Various flavonoids have been found useful for management of diabetes, obesity, metabolic diseases, cancers, bone health, gut health, renal functions, liver diseases, neurological diseases, viral infections, overall immunity and other health improvements. Scientists throughout the world are studying the mechanisms of action of flavonoids against different diseases. Besides these mentioned complications, flavonoids provide overall health benefits. Flavonoids have potential to be functional food of importance in the near future.

Furthermore, dietary flavonoids have been found useful in protecting against the COVID-19 disease and its associated symptoms. Some flavonoids have been found to inhibit the replication of coronaviruses. A balanced, flavonoid-rich diet has proven effective in the management of various diseases. Recent advances in the field have important insights about the role of specific flavonoids against certain diseases, dose dependent activity, mechanism of action, combination therapy, clinical trials, and other related aspects. All these aspects require more elaborative research.

In this current book, we have focused on the dietary flavonoids and their biological activity along with their therapeutic potential. The chapters describe the flavonoids that have an important role in human health management. In the first chapter, general perspectives of flavonoids have been presented. The chapter discusses the major structure, biosynthesis and general mechanism of action and health benefits of various flavonoids. Chapter Two focuses on the dietary flavonoid quercetin. It is one of the important flavonoids, known for its antioxidant, antimicrobial and other biological properties. It is also well known for its therapeutic potential. It may be a molecule of interest in the prevention and management of various diseases and disorders. This chapter also focuses on sources of quercetin. In Chapter Three, citrus flavonoid naringin has been described. Naringin is a plant-based, naturally occurring flavonoid. Citrus fruits are a natural source of the flavonoid naringin. It has been found effective against various diseases. This chapter focuses on naringin, its biological properties and therapeutic potential. Chapter Four discusses silymarin, a flavonoid with diverse medical properties. Milk thistle (*Silybum marianum*) is a medicinal plant that has been used for thousands of years as a remedy for a variety of ailments. The main component of *S. marianum* fruit extract is silymarin, a flavonolignan also known as silybin. Among the several potential biological activities, hepatoprotective activity is the main thing for which silymarin compound is used worldwide. Besides hepatoprotection, it is also beneficial for other disorders such as renal protection, hypolipidemic, cardiovascular protection, antioxidant. In neurological disorders, it is useful in the prevention of insulin resistance, cancer and Alzheimer's. It is also used as a food remedy and dietary supplement. This chapter aims to explain the various therapeutic potentials of silymarin along with its chemistry and mechanism of action. In Chapter Five, luteolin has been discussed. Luteolin is one of the well-known flavonoids abundantly present in a plant-based diet. It has been reported for its anti-cancer, antioxidant, anti-apoptotic, and anti-inflammatory properties. Luteolin has also been reported for its protective nature against viruses. It may be a promising molecule of interest against COVID-19.

In Chapter Six, the dietary flavonoid epigallocatechin has been discussed. Epigallocatechin is one of the important flavonoids known for its health benefits. Currently, green tea is of great attraction due to the presence of flavonoids and catechins. The catechins present in these teas have grabbed a lot attention due to their antimicrobial, anti-inflammatory, anticancer, anti-obesity, anti-glycation, and neuro- and cardioprotective effects. The chapter focuses on the potential of epigallocatechin as antioxidant, antimicrobial, anticancer, anti-inflammatory agent, antiaging agent as well as its role in preventing cardiovascular diseases. Genistein has been described in Chapter Seven. Genistein has exhibited various biological effects, such as improvement of glucose metabolism, anti-tumour activity (inhibition of cell proliferation, regulation of the cell cycle, induction of apoptosis), reduction of peri-menopausal and postmenopausal hot flashes, impairment of angiogenesis in both hormone-related and hormone-unrelated cancer cells, and modulation of antioxidant effects. Furthermore, epidemiological and clinical studies reported a number of health benefits of genistein in several chronic diseases, such as diabetes, osteoporosis, and cardiovascular disease, and help in the enhancement of distinctive menopausal symptoms, such as depression and anxiety. Even though the biological effects are more promising having certain restrictions, such as effects on target organs, low bioavailability and biological estrogenic activity, the clinical bids of genistein have been limited to some extent. Besides, few studies reported that molecular structure modification of genistein may eliminate the activity of estrogen and target organs. In the present review, we summarize the latent benefits of genistein on symptoms of menopause and diseases related to menopause like osteoporosis, anxiety, obesity, cardiovascular, diabetes, breast cancer and depression. Chapter Eight focuses on the flavonoid "Delphinidin." Delphinidin and its glycosides are water-soluble pigments, belonging to anthocyanidins, a subgroup of flavonoids. These are health-promoting candidatures for pharmaceutical and nutraceutical applications, as substantiated by exhibiting antioxidation, anti-inflammation, anti-microorganism, anti-diabetes, anti-obesity, cardiovascular protection, neuroprotection, and anticancer potentials. Additional investigation on the fundamental mechanisms of Del and its glycosides in mediating cellular processes with high specificity are still indispensable. In the present chapter, authors compiled the dietary sources, functional properties and health benefits of delphinidin and a glycosylated form of delphinidin.

Flavonoids are now well known for their health benefits, biological properties and therapeutic potential. Dietary-based intake of flavonoids is very important for human health. People throughout the world are now more conscious about their dietary intake of flavonoids. Also, people are using dietary flavonoids to maintain better health. Flavonoids are important, trending therapeutic agents that have immense potential to be used as functional food. In COVID-19 pandemic conditions, flavonoids have emerged as important natural anti-viral therapeutic agents. Flavonoids also helped people to build immunity and improve overall health. We hope this book will enhance the knowledge of life sciences (Biotechnology, Biochemistry, Pharmaceutical Technology) students (UG/PG), researchers, dieticians, and nutritionists about dietary flavonoids, and also be useful for persons working in related industries. This will also be useful for the students, researchers, industry persons and academicians engaged in the area of food technology, plant secondary metabolites, pharmacology, herbal formulations, and functional foods. This may also be a book of choice for common peoples having interest about the foods, dietary flavonoids, and health benefits of plant-based foods.

Chapter 1

A General Perspective of Flavonoids

Diwakar Aggarwal[1,*], Vasudha Datta[1] and Chadetrik Rout[2]

[1]Department of Biotechnology, Maharishi Markandeshwar (Deemed to be University), Mullana-Ambala, Haryana, India
[2]Department of Civil Engineering, Maharishi Markandeshwar (Deemed to be University), Mullana-Ambala, Haryana, India

Abstract

Flavonoids are a class of plant polyphenols with a wide range of chemical structures. They are typically found as secondary metabolites in photosynthetic plants. Common sources are fruits, vegetables, bark, roots, stems, flowers, tea and plants based processed food including wine. Plants show different colors in different parts because of flavonoids. They have a wide range of applications in nutraceuticals, pharmaceuticals, therapeutic products, and cosmetics, all of which have significant health benefits. . They have anti-oxidant, anti-inflammatory, anti-mutagenic, and anti-carcinogenic properties. Approximately 8000 to 10,000 distinct flavonoids have been identified to date. Out of numerous subclasses the major ones are, flavones, flavonols, flavanones, catechins or flavanols, anthocyanidins and isoflavones. Daily intake of flavonoids is recommended for healthy living, strong immunity and optimum cellular functions of the body.

Keywords: antioxidant, cancer, dietary flavonoids, polyphenols, secondary metabolites

* Corresponding Author's Email: diwakaraggarwal@yahoo.co.in.

In: Flavonoids
Editors: Nirmala Sehrawat, Diwakar Aggarwal, Sushil Kumar Upadhyay et al.
ISBN: 978-1-68507-953-6

1. Introduction

Flavonoids are diverse plant secondary metabolites also considered as plant pigments. They are commonly found in fruits, grains, vegetables, tea, and wine (Panche et al., 2016). They often display different colors in the different plant parts like in flowers, leaves etc. Actually, the word flavonoids originated from the Latin word flavus, meaning yellow color in nature. They are hydroxylated polyphenolic based plant metabolites and are frequently found attached to sugars (glycosides), making them water-soluble. Flavonoids can undergo further modifications, for example methylation by methyl-transferases and glycosylation by specific glycosyltransferases which alter their solubility, reactivity, and stability (Falcone -Ferreyra et al., 2012). Because of their diverse chemical structure and variety resulting from the attached substituents, they perform several important functions in plants. Primarily they work in the defense mechanism of the plant and defend the plant against various biotic and abiotic stress. Besides working in defense mechanism of plants they also help in attracting insects for pollination regulating cell growth and preventing plants from various microbial infections (Nijveldt et al., 2001).

Flavonoids were originally discovered in1930 by Albert Szent-Gyorgyi from orange and named as vitamin P as it was believed to be a vitamin responsible for curing scurvy along with vitamin C, further characterization revealed that the compound was not a vitamin but a flavonoid, named rutin (Kumar and Pandey 2013). This ignited the research on these novel compounds and till now approximately 800-10000 flavonoids have been identified and characterized from various natural resources (Khajuria et al., 2019). Later they were found to possess anti-oxidative activity, free radical scavenging ability, prevention of coronary heart disease, anti-atherosclerotic, hepatoprotective, anti-inflammatory and more recently anti-cancer properties. As a result, the pharmaceutical and healthcare sectors are paying close attention to flavonoids. Their anti-oxidant property is reported to be due to the presence of various functional hydroxyl groups in flavonoids structure and they act either by chelating metal ions and/or scavenging free radicals (Kumar et al., 2013). This metal chelating property helps in damaging target biomolecules. Moreover, flavonoids are reported to induce human protective enzyme systems, which in turn help to enhance immune repose to fight against both bacterial and viral infections and other major diseases related to cardiovascular and immune systems (Leopoldini et al., 2006). Moreover, flavonoids are also reported to show antiviral properties against viruses like herpes simplex, polio and pseudorabies (Arct and Pytkowska, 2008). Various studies have also demonstrated that flavonoids also act pro-oxidants (Van Acker et al., 1996, Cao et al., 1997, Agati et al., 2012).

2. Structure and Chemistry of Flavonoids

Chemically flavonoids are based upon a 15-carbon skeleton consisting of two phenyl rings (shown as A and B) linked via a heterocyclic ring (shown as C) with embedded oxygen (Figure 1). This is generally known as C6-C3-C6 framework and also considered as phenyl benzopyran functionality (Tsao 2010).

Further, the linkage of aromatic rings at different positions with respect to *benzopyrene* functionality helps to determine the further classes of flavonoids. These classes are flavonoids (2-phenylbenzopyrans), isoflavonoids (3-phenylbenzopyrans, neoflavonoids (4-phenylbenzopyrans). These classes are structurally and biogenetically same as they share a common chalcone precursor (Grotewold, 2006). Based on their chemical structure, flavonoids are classified into 12 major subclasses, six of which are of dietary importance, these are anthocyanidins, flavan-3-ols, flavonols, flavones, flavanones, and isoflavones. Furthermore, glycosylated flavonols are found most widely distributed flavonoids in the diet (Havsteen 2002). Major six subclasses classes of dietary flavonoids along with their source, structure and representative flavonoid are summarized in Table 1 below.

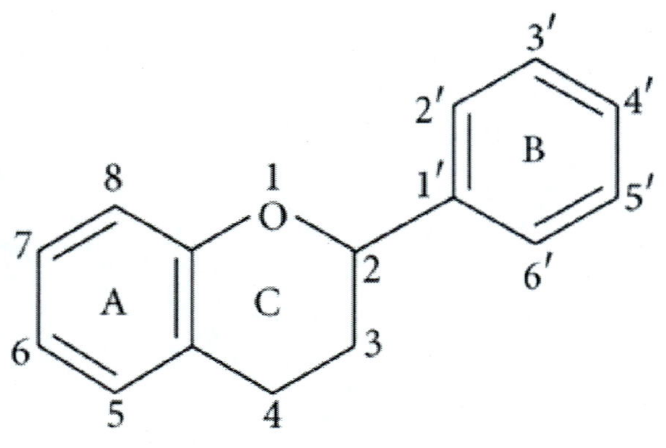

Figure 1. Basic structure of flavonoids.

Table 1. Major subclasses of flavonoids along with their structure, source and examples

Subclass of flavonoid	Basic Structure	Food sources	Flavonoid example
Flavanols	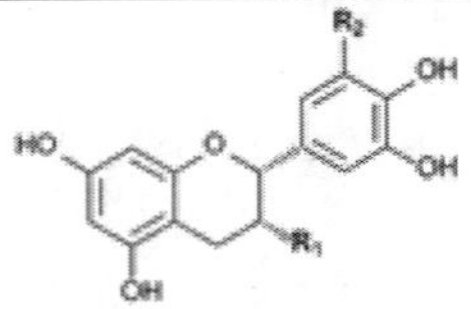	Apples, hops, tea, beer, wine, fruit juice	Epigallocatechin gallate, Catechins
Flavanones	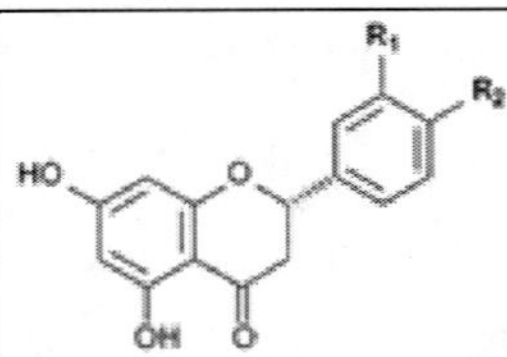	Citrus fruits and grapefruits,	Hespridin, Naringenin
Flavones	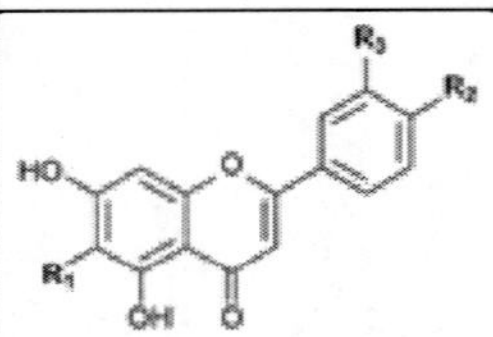	Herbs, cereals, fruits, vegetables, flowers	Apigenin¸ Luteolin
Flavonols	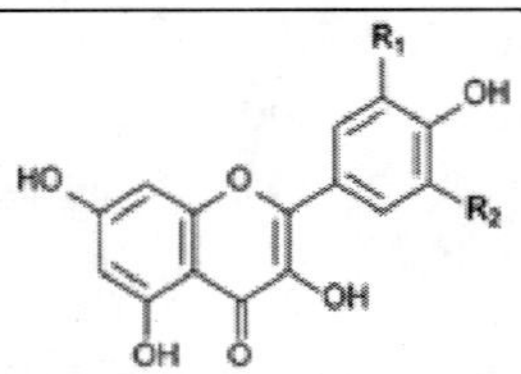	Onions, cherries, broccoli, kale, tomatoes, berries, buckwheat	Kaempferol, Quercetin
Anthocyanin	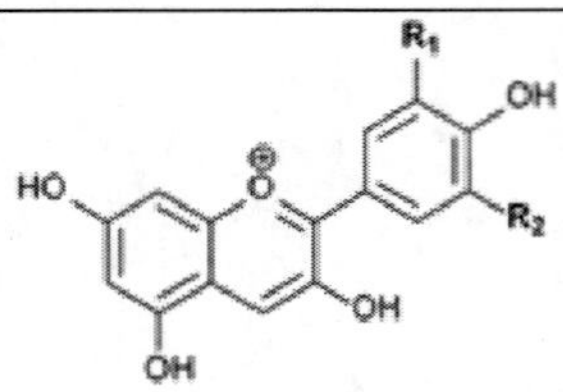	fruits, vegetables, nuts, olives and cereals	Cyanidin, Delphinidin
Isoflavonoids	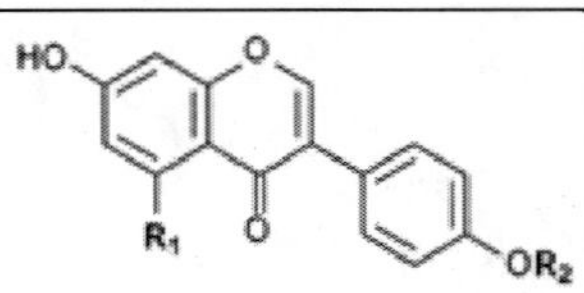	Legumes like lentils, beans, peas, Soy and its products	Genistein, Daidzein

Dietary flavonoids differ in the arrangements of hydroxyl, methoxy and glycosidic sidegroups, and in the conjugation between the A- and B- rings. During metabolism, hydroxyl groups are added, methylated, sulfated or glucuronidated. In food, flavonoids exist primarily as 3-O-glycosides and polymers. Moreover, flavonoids are reported to occur as aglycones, glycosides, and methylated derivatives as their basic structure are aglycone as shown in Figure 1. Condensation of six-member ring with the benzene ring is either a γ-pyrone (flavonols and flavanones) or its dihydro derivative (flavonols and flavanones). Flavonols differ from flavanones by hydroxyl group at the 3-position and a C2-C3 double bond (Narayana et al., 2001). As already discussed flavonoids are aglycones; usually these compounds exist as glycosides and but sometimes they also occur as acyl glycosides in fruits and vegetables. Furthermore, the attachment of sugar units to basic structure makes them *C*-Glycosyl flavonoids, through carbon–carbon bonding, these types are more frequently found in plants, as compared with fruits and vegetables. Some other derivatives of flavonoids like acylated and methylated flavonoids along with flavonoid sulfates have also been reported but in very low frequency and concentration as compared with others (Anderson and Markham, 2006).

3. Biosynthesis of Flavonoids

Flavonoids have been reported to be synthesized following two basic pathways i.e., the shikimic acid pathway and the malonic pathway. The former is responsible for the synthesis of flavonoids as plant phenolics in plants and later is more prominent in fungi and bacteria (Vogt, 2010). Basically various carbohydrate precursors are converted into amino acids phenylalanine and tyrosine in the shikimic acid pathway with shikimic acid as an intermediate. Shikimic acid synthesis is affected by commonly used herbicide glyphosate which blocks its synthesis. Animals are not able to synthesize flavonoids as they don't have the ability to synthesize the amino acids (i.e., phenylalanine, tyrosine, and tryptophan), and they rely upon plants to meet their demand for these essential amino acids. As flavonoids are derived from phenylalanine and tyrosine and therefore, they have an aromatic ring and a three-carbon side chain in their structure. As compared with tyrosine, phenylalanine is the preferred substrate for synthesis of flavonoids in plants (Aherne and Obrien, 2002).

4. Mechanism of Action of Flavonoids

A large number of *in vitro* studies have confirmed that flavonoids can inhibit and can also induce a variety of enzyme systems in humans and animals (Kandaswami and Middleton, 1994). The enzymes affected are known to be involved in various important pathways of the bodythat are known to regulate cell division and proliferation, platelet aggregation, detoxification, and various immunomodulation and inflammatory responses. So that's why flavonoids are found to be effective in various diseases related to immune system like cancer and homeostasis in cell systems etc. (Robak and Gryglewski, 1996). Further, they are also associated with a decreased risk of age-related diseases because of the above properties (Pollastri and Tattini, 2011).

As flavonoids are antioxidants by their chemical nature are, they exhibit a wide array of antioxidant properties, which in turn help in protecting the tissues against oxygen free radicals and lipid peroxidation. Oxygen free radicals and lipid peroxidation are further reported to be involved in various pathological conditions such as atherosclerosis, cancer, and chronic inflammation (Tuli et al., 2021), and therefore flavonoids are used for prevention and cure of these diseases. Now a day's flavonoids are most commonly used as agents to cure and prevent cancer, major mechanisms responsible for prevention and cure of cancer are down regulation of pro apoptotic genes, cell cycle arrest, up-regulation of anti-apoptotic genes and inhibition of expression of Ras proteins (Tuli et al., 2021). Through their activity of scavenging the common free radicals and chelating metals that generate ROS, they provide stress protection via the Fenton reaction (Williams et al., 2004). They are also reported to provide resistance to metal toxicity in plants (Kidd et al., 2001).

5. Dietary Sources of Flavonoids

As the world is now aware of the amazing health benefits of flavonoids, therefore foods rich in flavonoids are generally considered superfoods (Liu, 2013). The Primary source of flavonoids are foods of plant origin like tea, fruit, vegetables, grains, legumes, nuts and wine, with tea and wine being the primary dietary sources of flavonoids in eastern and western societies of the world (Pallauf et al., 2017). Besides these, leafy vegetables, onions, apples, berries, cherries, soybeans, and citrus fruits are considered important sources of dietary flavonoids. Even beer is reported to have a considerable amount of flavonoid xanthohumol (Aggarwal et al., 2021). Individual intake, amount and type of flavonoid may vary considerably according to food. Various database including USDA provides information on the flavonoid content of the various foods (Bhagwat et al., 2014). If we consider various subclasses of flavonoids, flavonols are reported to constitute the most abundant flavonoids in foods, followed by flavanones that occur in citrus fruits and flavones in herbs. Anthocyanins and isoflavones are maximally reported from strawberries, other berries and soy foods respectively (Aherne and Obrien 2002), while catechins are reported from tea (Yao et al., 2004).

Recommended intake of flavonoids either individually or in combination with other supplements like synthetic antioxidants, and other vitamins can help in reducing the risks of cancer, cardiovascular related deaths and other major immune system related diseases (Herrera et al., 2009). With regular intake of fruits and vegetables one can meet the daily intake of these compounds i.e., ranging from 50 and 800 mg/day (Spencer et al., 2009). For healthy living, strong immunity and optimum cellular functions of the body it is important to consume the recommended daily intake of these compounds (Procházková et al., 2011).

Conclusion

In the last few years flavonoids have acquired a wider range of interest due to their immense health benefits. Every year, new flavonoids are discovered with medicinal promise in the prevention and treatment of various diseases. The fact that these beneficial substances may be found in a wide range of fruits and vegetables and have become commonplace in human and animal diets piqued researchers' and the general public's curiosity.

References

Agati, G., Azzarello, E., Pollastri, S. & Tattini, M. (2012). Flavonoids as antioxidants in plants: location and functional significance. *Plant Science*, 196, 67-76.

Aggarwal, D., Upadhyay, S. K., Singh, R. & Tuli, H. S. (2021). Recent patents on therapeutic activities of xanthohumol: a prenylatedchalconoid from hops (*Humulus lupulus* L.). *Pharmaceutical Patent Analyst*, 10, 37-49.

Aherne, S. A. & O'Brien, N. M. (2002). Dietary flavonols: chemistry, food content, and metabolism. *Nutrition*, 18, 75-81.

Andersen, O. M. & Markham, K. R. (2006). *Flavonoids: chemistry, biochemistry and applications*. CRC press.

Arct, J. & Pytkowska, K. (2008). Flavonoids as components of biologically active cosmeceuticals. *Clinics in Dermatology*, 26, 347-357.

Bhagwat, S., Haytowitz, D. B. & Holden, J. M. (2014). USDA database for the flavonoid content of selected foods, *Release* 3.1. US Department of Agriculture: Beltsville, MD, USA.

Cao, G., Sofic, E. & Prior, R. L. (1997). Antioxidant and prooxidant behavior of flavonoids: structure-activity relationships. *Free Radical Biology and Medicine*, 22, 749-760.

Falcone-Ferreyra, M. L., Rius, S. & Casati, P. (2012). Flavonoids: biosynthesis, biological functions, and biotechnological applications. *Frontiers in Plant Science*, 3, 222.

Grotewold, E. (Ed.). (2006). *The science of flavonoids* (pp. 1-274). New York: Springer.

Havsteen, B. H. (2002). The Biochemistry and medical significance of the flavonoids. *Pharmacology & Therapeutics*, 96, 67-202.

Herrera, E., Jiménez, R., Aruoma, O. I., Hercberg, S., Sánchez-García, I. & Fraga, C. (2009). Aspects of antioxidant foods and supplements in health and disease. *Nutrition Reviews*, 67, S140-S144.

Kandaswami, C. & Middleton, E. (1994) Free Radical Scavenging and Antioxidant Activity of Plant Flavonoids. In: Armstrong D. (eds) *Free Radicals in Diagnostic Medicine. Advances in Experimental Medicine and Biology*, Vol. 366. Springer, Boston, MA. https://doi.org/10.1007/978-1-4615-1833-4_25.

Khajuria, R., Singh, S. & Bahl, A. (2019). General introduction and sources of flavonoids. In *Current aspects of flavonoids: Their role in cancer treatment* (pp. 1-7). Springer, Singapore.

Kidd, P. S., Llugany, M., Poschenrieder, C. H., Gunse, B. & Barcelo, J. (2001). The role of root exudates in aluminium resistance and silicon-induced amelioration of aluminium toxicity in three varieties of maize (*Zea mays* L.). *Journal of Experimental Botany*, 52, 1339-1352.

Kumar, S. & Pandey, A. K. (2013). Chemistry and biological activities of flavonoids: an overview. *The Scientific World Journal*, 2013, https://doi.org/10.1155/2013/162750.

Kumar, S., Mishra, A. & Pandey, A. K. (2013). Antioxidant mediated protective effect of *Parthenium hysterophorus* against oxidative damage using *in vitro* models. *BMC Complementary and Alternative Medicine*, 13, 1-9.

Leopoldini, M., Russo, N., Chiodo, S. & Toscano, M. (2006). Iron chelation by the powerful antioxidant flavonoid quercetin. *Journal of Agricultural and Food Chemistry*, 54, 6343-6351.

Liu, R. H. (2013). Health-promoting components of fruits and vegetables in the diet. *Advances in Nutrition*, 4, 384S-392S.

Narayana, K. R., Reddy, M. S., Chaluvadi, M. R. & Krishna, D. R. (2001). Bioflavonoids classification, pharmacological, biochemical effects and therapeutic potential. *Indian Journal of Pharmacology*, 33, 2-16.

Nijveldt, R. J., Van Nood, E. L. S., Van Hoorn, D. E., Boelens, P. G., Van Norren, K. & Van Leeuwen, P. A. (2001). Flavonoids: a review of probable mechanisms of action and potential applications. *The American Journal of Clinical Nutrition*, 74(4), 418-425.

Pallauf, K., Duckstein, N. & Rimbach, G. (2017). A literature review of flavonoids and lifespan in model organisms. *Proceedings of the Nutrition Society*, 76, 145-162.

Panche, A. N., Diwan, A. D. & Chandra, S. R. (2016). Flavonoids: an overview. *Journal of Nutritional Science*, 5, 47-62.

Pollastri, S. & Tattini, M. (2011). Flavonols: old compound for old roles. *Annals of Botany*, 108, 1225-1233.

Procházková, D., Boušová, I. &Wilhelmová, N. (2011). Antioxidant and prooxidant properties of flavonoids. *Fitoterapia*, 82, 513-523.

Robak, J. & Gryglewski, R. J. (1996). Bioactivity of flavonoids. *Polish Journal of Pharmacology*, 48, 555-564.

Spencer, E. H., Ferdowsian, H. R. & Barnard, N. D. (2009). Diet and acne: a review of the evidence. *International Journal of Dermatology*, 48, 339-347.

Tsao, R. (2010). Chemistry and biochemistry of dietary polyphenols. *Nutrients*, 2, 1231-1246.

Tuli, H. S., Aggarwal, V., Tuorkey, M., Aggarwal, D., Parashar, N. C., Varol, M., Savla, R., Kaur, G., Mittal, S. & Sak, K. (2021a). Emodin: A metabolite that exhibits anti-neoplastic activities by modulating multiple oncogenic targets. *Toxicology in Vitro*, 73, 105142.

Tuli, H. S., Joshi, R., Aggarwal, D., Kaur, G., Kaur, J., Kumar, M., Parashar, N. C., Khan, M. A. & Sak, K. (2021). Molecular mechanisms underlying chemopreventive potential of butein: Current trends and future perspectives. *Chemico-Biological Interactions*, 350, 109699.

Van Acker, S. A., Tromp, M. N., Griffioen, D. H., Van Bennekom, W. P., Van Der Vijgh, W. J. & Bast, A. (1996). Structural aspects of antioxidant activity of flavonoids. *Free Radical Biology and Medicine*, 20, 331-342.

Vogt, T. (2010). Phenylpropanoid biosynthesis. *Molecular Plant*, 3, 2-20.

Williams, R., Spencer, J. and RiceEvans, C. (2004). Flavonoids: antioxidants or signaling molecules? *Free Radical Biology & Medicine*. 36, 838-849.

Yao, L. H., Jiang, Y. M., Shi, J., Tomas-Barberan, F. A., Datta, N., Singanusong, R. & Chen, S. S. (2004). Flavonoids in food and their health benefits. *Plant Foods for Human Nutrition*, 59, 113-122.

Chapter 2

Quercetin: Dietary Sources, Biological Properties and Therapeutic Potential

Sachin Kumar[1,*], Nirmala Sehrawat[2], Mukesh Yadav[2], Anil Kumar Sharma[2], Sneh Lata[3], Amit Kumar[4], Sunil Kumar[2] and Ashwani Kumar[5]

[1]Deptartment of Bioinformatics, Janta Vedic College, Baraut- Baghpat, Uttar Pradesh, India
[2]Department of Biotechnology, Maharishi Markandeshwar (Deemed to be University), Mullana-Ambala, India
[3]University Institute of Biotechnology, Chandigarh University, Gharuan, Punjab, India
[4]Department of Biotechnology, School of Engineering and Technology, Sharda University, Greater Noida, India
[5]Department of Biotechnology, Chaudhary Bansi Lal University, Bhiwani, Haryana, India

Abstract

Flavonoids are now well known for their health benefits, biological properties and therapeutic potential. Plants are the major source of the flavonoids. Dietary based intake of flavonoids is very important for human health. Quercetin is one of the important flavonoids. It is well known for its anti-oxidant, antimicrobial and other biological properties. Also, it is well known for its therapeutic potential. It may be a molecule of interest in prevention and management of various diseases and disorders. This chapter focuses on sources of quercetin, its biological properties and therapeutic potential.

---

* Corresponding Author's Email: sachinsuryan@gmail.com.

In: Flavonoids
Editors: Nirmala Sehrawat, Diwakar Aggarwal, Sushil Kumar Upadhyay et al.
ISBN: 978-1-68507-953-6

Keywords: quercetin, flavonoid, dietary sources, health benefits

1. Introduction

Flavonoids, or commonly known as "bioflavonoids," are a class of polyphenolic secondary metabolites widely distributed in nature. The term flavonoids originated from Latin word "flavus," that means yellow, because of their colour in nature. The first flavonoid was isolated and identified by Szent-Gyorgyi, around 1936 in the form of vitamin P or citrin which raised a controversy as to whether "vitamin P" met all the classical requirements of being a vitamin even though it was nutritionally and therapeutically useful. The term "bioflavonoid" was introduced to designate the flavonoids having biological activity. The term was first suggested by B.L. Oser, while it was used in the literature first time by EF Bryant. Therefore, vitamin P was the first crude citrus bioflavonoid investigated by Szent-Gyorgyi (Baier, 1955).

Flavonoids are abundant in plants in which they execute several functions e.g., produce colour to attract pollinating insects, works as chemical messengers, also required for nitrogen fixation, UV filtration and cell cycle inhibition etc. Flavonoids also help in development of symbiotic relationship between rhizobia and some vegetables such as clover, beans and peas. In presence of flavonoids, the rhizobia produce "Nod factors" which induces root nodule formation. Some flavonoids also protect plants against the disease by inhibiting certain spores. Flavonoids are abundant in plants and thereby are the most common type of polyphenolic compound found in the human diet. Bioflavonoids are believed to provide health benefits through cell signalling pathways and antioxidant properties.

Being important antioxidants, flavonoids promote several health effects. With their low toxicity in comparison to other plant compounds, and their abundance, flavonoids can be ingested by animals in a massive quantity, including humans. Apart from their antioxidant activity flavonoids have anti-viral, anti-inflammatory, anti-cancer, and anti-allergic effects.

Flavonoids are polyphenolic molecules having general structure of a 15-carbon skeleton (Flavone), comprises of two phenyl rings (A and B) and a heterocyclic ring (Figure 1).

Flavonoids derived from nature have various substitutions, most frequently in the 3, 5, 7, 3' and 4' positions. Based on the kind of substitutions they have bioflavonoids are classified into different subgroups namely anthocyanidins, flavons, flavonols, anthoxanthins, and, flavanones (Table 1,

Figure 2). The flavone that has a substitution of H in the 3rd position of pyrone ring with an OH group is termed flavonol. When the double bond at 3rd position in the original flavone is saturated with H atoms, the chemical compound is termed flavanone. Anthocyanidin is the class of compound resulted from a drastic reduction of carbonyl of pyrone ring, as with magnesium in acid alcohol, subsequent saturation of this ring with hydrogen gives catechin.

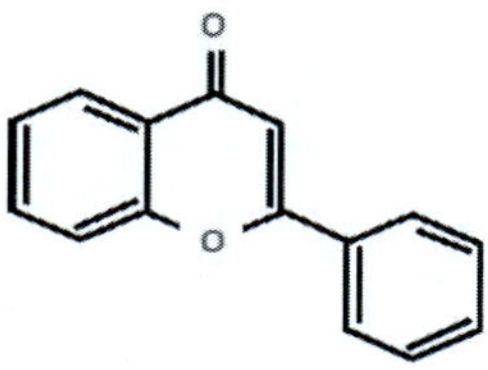

Figure 1. Chemical structure of flavone backbone (2-Phenyl-1,4-benzopyrone).

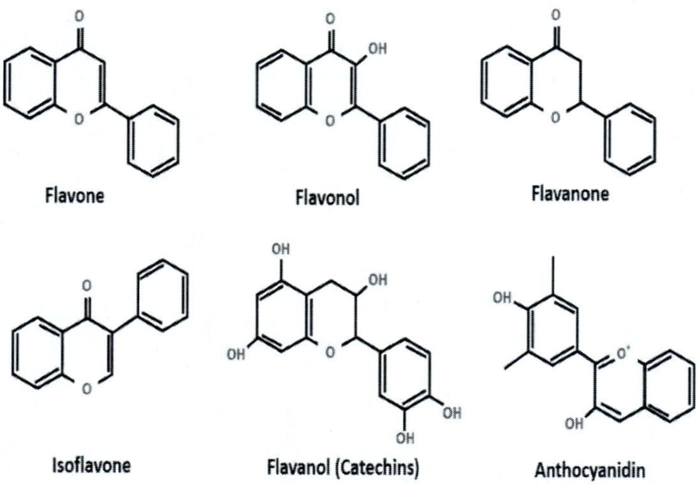

Figure 2. Structure of different classes of flavonoids.

Quercetin is considered as a member of flavonol subclass of flavonoids; the name is derived from quercetum (oak forest), after Quercus, and has been used since 1857. Quercetin is a polar auxin transport inhibitor, occurring naturally in plants (Fischer et al., 1997). Quercetin has an OH group attached at positions 3, 5, 7, 3', and 4', the International Union of Pure and Applied Chemistry (IUPAC) nomenclature for quercetin is 2-(3,4-dihydroxyphenyl)-3,5,7-trihydroxychromen-4-one. Common forms of quercetin are shown in Figure 3.

Table 1. Main groups of flavonoids, compounds and food sources (Lakhanpal and Kumar, 2007)

Class of Flavonoid	Example Compounds	Food sources
Flavonols	Quercetin, Kaempferol Myricetin, Isorhamnetin Querctagetin	Yellow onion, Curly kale, Leek, Cherry tomato, Broccoli, Apple, Green and black tea, Black grapes, Blueberry.
Flavanones	Naringenin, Eriodictyol Dihydroquercetin Dihydrofisetin, Hesperetin, Dihydrobinetin	Orange juice, Grapefruit juice, Lemon juice.
Flavanols	Silibinin, Silymarin Taxifolin, Pinobanksin	Cocoa, Cocoa beverages, Chocolates.
Flavones	Tangeretin, Heptamethoxyflavone Nobiletin, Sinensetin, Quercetogetin, Chrysin, Apegenin, Luteolin, Disometin, Tricetin	Parsley, Celery, Capsicum pepper.
Anthocyanins	Cyanidin, Delphinidin Malvidin, Pelargonidin Peonidin, Petunidin	Blue berry, Blackcurrant, Black grapes, Cherry, Rhubarb, Plum, Strawberry, Red wine, Red cabbage.
Isoflavones	Daidzein, Genistein, Glycitein	Soy cheese, Soy flour, Soy bean, Tofu.
Catechins (Flavans) (Proanthocyanidins)	(+) Catechin, Gallocatechin, (-) Epicatechin, Epigallocatechin Epicatechin 3-gallate Epigallocatechin 3-gallate	Chocolate, Beans, Apricot, Cherry, Grapes, Peach, Red wine, Cider, Green tea, Black tea, Blackberry.

Quercetin ($C_{15}H_{10}O_7$) is itself an aglycone, it do not have any attached sugar. A quercetin glycoside (the most common form) is formed when a glycosyl group (a sugar such as glucose, rhamnose, or rutinose) is attached by replacing one of the OH groups (commonly at position 3). Quercetin itself is almost insoluble in cold water, but it is soluble in alcohol and lipids. The attached glycosyl group alters the properties of quercetin i.e., the solubility, absorption, and in vivo effects. Presence of a glycosyl group in quercetin glycoside results in increased water solubility in comparison to quercetin aglycone (Ross and Kasum, 2002; Hollman et al., 1999).

Normally, the term quercetin should be used to describe the aglycone only; though, the name is sporadically used to refer to quercetin type molecules, including its glycosides.

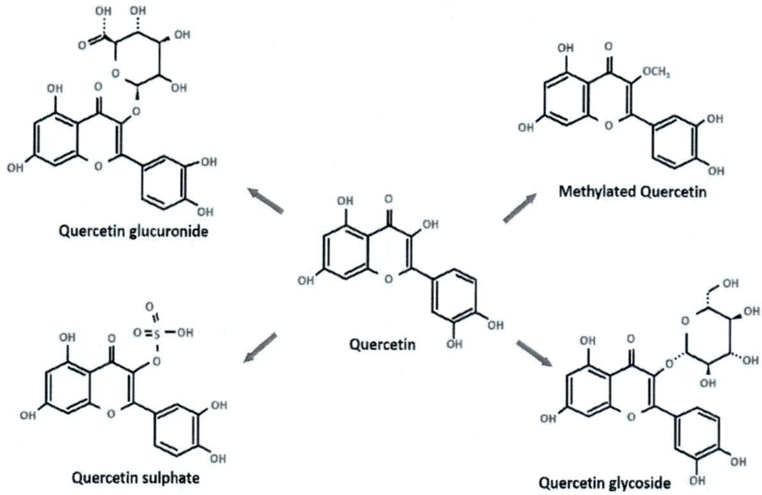

Figure 3. Molecular structure quercetin, and its different forms i.e., quercetin sulphate, quercetin glycoside, methylated quercetin and quercetin glucuronide.

Table 2. Amount of Quercetin in selected food (Mangels et al., 1993)

Food Source	Quercetin (mg/100g of Food)
Cocoa powder, Unsweetened	20.1
Cranberries, Raw	14.0
Lingonberries, Raw	11.3
Kale, Raw	5.1
Celery, Raw	3.5
Broccoli, Raw	2.8
Lettuce, Looseleaf, Raw	2.0
Tomatoes, Red ripe, Raw	0.5
Carrots, Raw	0.4

Plants are the richest source for bioflavonoids (Anand-David et al., 2016). Quercetin is found in a variety of fruits and vegetables. Quercetin-type flavonols (primarily as quercetin glycosides), the most abundant of the flavonoid molecules, are widely distributed in the plant kingdom. They are found in a variety of foods including apples, berries, "Brassica" vegetables, capers, grapes, onions, shallots, tea, and tomatoes, as well as many seeds, nuts, flowers, barks, and leaves. Studies have been carried out to analyse flavonoid content in different food sources at various times by various researchers. Quercetin content in selected food is stated in Table 2 and Table3 as cited in different studies carried out at different times. Flavonoids content among 62 edible tropical plants was studied by Miean and Mohamed in 2001.

Table 3. Mean value of Quercetin in selected food that has quercetin content of 10mg or more than 10 mg/100g of food part studied (Bhagwat et al., 2013)

Food Source (with Analysed part, condition)	Quercetin (mg/100g of Food)	Food Source (with Analysed part, condition)	Quercetin (mg/100g of Food)	Food Source (with Analysed part, condition)	Quercetin (mg/100g of Food)
Capers, raw	233.84	Lovage, leaves, raw	170.00	Capers, canned (*Capparis spinosa*)	172.55
Juniper berries, ripe (*Juniperus communis*)	46.61	Juniper berries, green, unripe (*Juniperus communis*)	42.81	Coriander (cilantro) leaves, raw (*Coriandrum sativum*)	52.90
Lingonberries (cowberries), raw	13.30	Plums, black diamond, with peel, raw	12.45	Service (Saskatoon) berries (*Amelanchier canadensis*)	16.64
Annual saw-thistle, leaves	16.00	Asparagus, cooked, boiled, drained	15.16	Asparagus, raw (*Asparagus officinalis*)	13.98
Corn poppy, leaves	26.30	Dock, raw (*Rumex* spp.)	86.20		
Drumstick (horseradish tree) leaves, raw (Moringa oleifera)	16.65	Kale, raw (*Brassica oleracea* (Acephala Group))	22.58	Okra, raw (*Abelmoschus esculentus*)	20.97
Fennel, leaves, raw	48.80	Hartwort, leaves	29.30	Nalta jute, raw	23.53
Onions, cooked, boiled, drained, without salt	24.36	Onions, Raw	20.30	Onions, red, raw	39.21
Onions, spring, red, bulb	30.60	Onion white, pan- fried	26.90	Peppers, ancho	27.60
Peppers, hot, yellow wax, raw	50.63	Peppers, serrano,raw (*Capsicum anuum*)	15.98	Radish leaves, raw	70.37

Food Source (with Analysed part, condition)	Quercetin (mg/100g of Food)	Food Source (with Analysed part, condition)	Quercetin (mg/100g of Food)	Food Source (with Analysed part, condition)	Quercetin (mg/100g of Food)
Rocket, wild, raw (*Diplotaxis tenuifolia*)	66.19	Sweet potato leaves, raw (*Ipomoea batatas*)	16.94	Watercress, raw (*Nasturtium officinale*)	29.99
Chia seeds, raw	18.42	Bee Pollen	20.95	Buckwheat, bran	14.90
Carob fiber (Caromax)	58.13	Carob flour (*Ceratonia siliqua*)	38.78	Juice, cranberry, raw	16.41
Apple, skin only	19.36	Oregano, Mexican, dried	42.00	Chokeberry, raw	18.53
Cowpeas, black seed cultivar, mature seeds, raw (*Vigna unguiculata* subsp. Sinensis)	17.22	Dill weed, fresh (*Anethum graveolens*)	55.15	Buckwheat (*Fagopyrum esculentum* Moench)	15.38
Greek greens pie (prepared from wild greens)	12.40	Goji berry (wolfberry), dried	13.60	Cherries, sour, powder	17.44
Bog whortleberries, wild, frozen	17.70	Blueberries, rabbiteye, raw (*Vaccinium* spp.)	14.42	Elderberries, raw (*Sambucus* spp.)	26.77
Juice concentrate, elderberry	108.16	Juice concentrate, Chokeberry	68.17	Juice concentrate, black currant	22.85
Cranberries, raw (*Vaccinium macrocarpon*)	14.84	Radicchio, raw (*Cichorium intybus*)	31.51	Cocoa, dry powder, unsweetened	10.00

The total flavonoids contents were found highest in onion leaves (1497 mg/Kg Quercetin, 832 mg/kg Kaempferol and 391 mg/kg Luteolin). As per the Release 3.1 by USDA(Bhagwat et al., 2013)for the flavonoid content of selected foods, the highest concentration is 234 mg/100 g of edible portion in capers (raw), while the lowest concentration is 2 mg/100 g of edible portion in black or green tea (*Camellia sinensis*) among the determined food (Table2 and Table 3).

Table 4. List of plant species that have quercetin as one of the most substantial bioflavonoid compounds (Batiha et al., 2020)

1. *Foeniculum vulgare*	2. *Curcuma domestica valeton*	3. *Santalum album*
4. *Mangifera indica*	5. *Daucus carota*	6. *Emblica officinalis*
7. *Momordica charantia*	8. *Ocimum sanctum*	9. *Psoralea corylifolia*
10. *Solanum nigrum*	11. *Glycyrrhiza glabra*	12. *Morua alba*
13. *Allium fistulosum*	14. *Allium cepa*	15. *Calamus scipionum*
16. *Centella asiatica*	17. *Hypericum hircinum*	18. *Hypericum perforatum*
19. *Brassica oleracea var. italica*	20. *Brassica oleracea var. sabellica*	21. *Coriandrum sativum*
22. *Nasturtium officinale*	23. *Asparagus officinalis*	24. *Capparis spinosa*
25. *Prunus avium*	26. *Malus domestica*	27. *Vaccinium oxycoccus*
28. *Cuscuta reflexa*	29. *Camellia sinensis*	30. *Lactuca sativa*
31. *Withania somnifera*	32. *Moringa oleifera*	33. *Prunus domestica*
34. *Swertia chirayita*	35. *Apium graveolens*	36. *Solanum Lycopersicum*

Table 5. Estimated daily intake of quercetin by the people of different countries, their main food sources

Daily intake of flavonol, quercetin	Area, Country	Main food sources	Reference
Quercetin intake of 4.37 mg /day	Suihua area of northern China	Apple, Potatoes, Lettuce and Oranges	Sun et al., 2015
Quercetin intake of 4.43 mg /day	Harbin, China	Apple, Potatoes, Celery, Eggplant and Actinidia	Zhang et al., 2010
Flavonol intake of 13 mg /day (quercetin represents 75% of that)	Adults in USA	Onion, Apples and Tea	Sampson et al., 2002
Quercetin intake of 16.2 mg /day (average)	Japan	Onion, Green Tea, Asparagus, Green pepper, Tomatoes and Red leaf lettuce	Nishimuro et al.,2015
Quercetin intake of 18.48 mg /day (average)	Spain	Tea, Citrus fruits and juice, Beers and Ales, Wines, Melon, Apples, Onion, Berries and Bananas	Zamora-Ros et al., 2010

Plant species listed in Table 4 are among important source for quercetin. Quercetin is one of the most substantial bioflavonoid compounds found in fruits, grains, and vegetables for more than 30 plant species.

Dietary intake of quercetin has been found different in different countries typically depending on the consumption of fruits and vegetables and the intake of tea. The estimated daily intake of quercetin among different countries has been summarized in Table 5.

2. Biological Properties

Quercetin is a widely available flavonoid, which is synthesized by plants as a part of their defensive response to the environment e.g., in response to UV radiation and lipid peroxidation (Mohle et al., 1985; Mariani et al., 2008). Some of the important biological properties of quercetin are as mentioned below.

2.1. Antioxidant Properties

Among the biological properties of quercetin, the very first property of quercetin is its antioxidant potential, the potential of quercetin to act as an antioxidant is its most well-known property. Quercetin appears to be one of the most effective flavonoids for protecting the body against reactive oxygen species, which are formed naturally or as a result of exogenous damage (De Groot, 1994; Grace, 1994). Lipid peroxidation tends to be one of the most critical mechanisms and sequences of events by which free radicals interact with cellular functions, ultimately leading to cell death. Living organisms have evolved antioxidant protection mechanisms to prevent cellular death caused by reactive oxygen species (Halliwell, 1995).

Antioxidants, both enzymatic and non-enzymatic, hold ROS/RNS levels in check and repair oxidative cellular harm. Superoxide dismutase (SOD), catalase (CAT), and glutathione peroxidase (GPx) are the main enzymes involved in the first line of defence in the neutralization of ROS/RNS. Radical scavenging antioxidants like vitamin C, vitamin A, and plant phytochemicals like quercetin, which inhibit oxidation chain initiation and prevent chain propagation, are the second line of defence. End of an oxidation chain caused by the reaction of two radicals is also included in second line of defence.

Repair and membrane reconstitution by repair and de novo enzymes (Lipases, proteases, DNA repair enzymes, and transferases) serves as the third line of defence (Bahorun et al., 2006).

2.2. Direct Radical Scavenging Action

Free radicals are produced in animal cells both accidentally as well as deliberately and because of being important biochemical intermediates, free radicals have been a serious concerned in a large number of human diseases (Wegener and Fintelmann, 1999; Ares and Outt, 1998). A protective effect in reperfusion ischemic tissue damage is shown to be exerted by Quercetin as a free radical scavenger (Santos et al., 1998; Halliwell, 1994; Fraga et al., 1987). Direct scavenging of free radicals is one of the various ways by which quercetin prevents the tissue injury induced by free radicals. LDL oxidation is inhibited in vitro by Quercetin (Kerry and Abbey, 1987).

2.3. Xanthine Oxidase Inhibition

The xanthine oxidase pathway has been implicated as a key pathway in tissue oxidative injury, especially following ischemia reperfusion (Santrueza et al., 1992). In the conversion of xanthine to uric acid, both xanthine dehydrogenase and xanthine oxidase are involved. Under physiological conditions, xanthine dehydrogenase is present, but under oxidative stress and ischemic conditions, its structure changes to xanthine oxidase. Quercetin appears to reduce oxidative injury by inhibiting xanthine oxidase activity (Shoskes, 1998; Chang et al., 1993; Lio et al., 1986).

2.4. Inhibition of Inducible Nitric Oxide Synthases

Interference of Quercetin with inducible nitric oxide synthase activity, results in a reduction in ischemia reperfusion injury (Shoskes, 1998). Macrophages, endothelial cells and several other different types of cells produce nitric oxide. Early release of nitric oxide is important in maintaining the dilatation of blood vessels (Huk et al., 1998), nitric oxide produced by inducible nitric oxide synthase at a high concentration in macrophages can result in oxidative damage. Peroxynitrite is produced from the reaction between free radicals and nitric oxide, which oxidize LDLs resulting in irreversible damage to cell membranes. Quercetin reduces this damage by scavenging free radicals and nitric oxide (Shutenko et al., 1990; van Acker et al., 1995).

2.5. Decreasing Leukocyte Immobilization

Another important mechanism is the immobilization and firm adhesion of leukocytes to the endothelial wall, which is responsible for the formation of oxygen-derived free radicals, as well as the release of cytotoxic oxidants and inflammatory mediators, as well as the activation of the complement system. Leukocytes travel easily around the endothelial walls under normal circumstances.

During ischemia and inflammation, however, numerous factors, primarily endothelial derived mediators and complement factors can trigger leukocyte adhesions to endothelial walls, immobilizing them and stimulating neutrophil degranulation. Oxidants and inflammatory mediators are released as a result, causing tissue damage. The number of immobilized leukocytes during reperfusion was documented to decrease after oral administration of distilled micronized flavonoids fraction, which could be linked to its anti-inflammatory mechanism (Friesenecker et al., 1994).

2.6. Interaction with Other Enzyme Systems

Quercetin interacts with a calcium regulatory protein termed calmodulin (Nishino et al., 1984). Calmodulin is a protein that carries calcium ions through cellular membranes, triggering a variety of cellular processes. The antioxidant quercetin tends to be a calmodulin antagonist. Quercetin acts as a membrane stabilizer at the cell membrane level through this mechanism (Busse et al., 1984). Calmodulin-dependent enzymes including ATPases and phospholipases present at the cell membrane are inhibited by quercetin, which affect membrane permeability (Haysteen, 1983). Furthermore, various cellular functions e.g., histamine secretions from mast cells that are regulated by calmodulin-dependent enzymes are also affected by quercetin.

Quercetin has been shown to suppress histamine secretion from mast cells in different tissues, as well as from basophils (Otsuka, 1995; Fox et al., 1988; Pearce et al., 1984; Middleton et al., 1981; Benett et al., 1981; Middleton and

Drzewiecki, 1985). Quercetin also inhibits phospholipases which catalyses the release of arachidonic acid.

Inflammatory prostaglandins and leukotrienes all use arachidonic acid as a key substrate. Quercetin also inhibits cyclooxygenase and Lipooxygenase, enzymes that catalyze the conversion of arachidonic acid to its metabolites (Middleton and Drzewiecki, 1985, Yoshimoto et al., 1983; Della et al., 1988).

It is beneficial to reduce the levels of these metabolites, as well as histamine, in order to preserve the natural comfort level of body tissue and structures. On endothelial cells, quercetin has also been shown to inhibit the action of adhesion molecules (Middleton and Suresh, 1995). Quercetin also chelates ions of transition metals like iron, which can cause oxygen free radicals to form (Afanasev et al., 1989; Ferrali et al., 1997). Another safeguard is the direct inhibition of lipid peroxidation (Sorata et al., 1984).

2.7. Modulation of Gene Expression

Recent research suggests that Quercetin's radical scavenger property is unlikely to be the sole reason for their neuroprotective abilities, and that their biological behaviour may be attributed to a broad range of cellular signaling events (Mandel et al., 2004). Quercetin's ability to interact with intracellular signaling pathways, such as the mitogen-activated protein kinase cascade, has gotten a lot of attention recently.

Quercetin's strong neurotoxic ability in primary cortical neurons may be due to complex and responsive interactions within the neuronal mitogen-activated protein kinase and Akt/protein kinase B (PKB) signaling cascades, both of which have been linked to neuronal apoptosis. Quercetin inhibited phosphorylation of both Akt/PKB and ERK, resulting in decreased BAD phosphorylation and high caspase-327 activation.

TNF-is one of the most significant proinflammatory cytokines involved in the pathogenesis of chronic inflammatory diseases, and it is influenced by oxidative stress (Calamia, 2003; Taylor et al., 2004). TNF- also stimulates the development of other cytokines, chemokines, and inflammatory mediators in the cell, as well as having antiviral and antimicrobial properties (Aggarwal, 2000; Aggarwal2001; Wajant et al., 2001).TNF-production and gene expression were significantly reduced by quercetin in a dose-dependent manner. The ability of flavonoids to modulate the immune response and have possible anti-inflammatory activity is demonstrated by a decrease in endogenous TNF- production in the presence of quercetin. TNF- has complex effects on the development, differentiation, and death of immune cells, in addition to its well-known pro-inflammatory function. TNF inhibition has been shown to be effective in the treatment of a variety of inflammatory diseases (Calamia, 2003). By inhibiting the activation of NF-, quercetin-induced suppression of TNF- will result in the stimulation of anti-inflammatory cytokines, and thus quercetin could be widely used as an anti-TNF- therapy.

Quercetin has anti-proliferative activity; the mechanisms of quercetin action may be by regulation of cell cycle and cell growth regulatory genes (Kaneuchi et al, 2003). Via the downregulation of EGF and cyclin D1, quercetin can inhibit the proliferation of Ishikawa cells (endometrial carcinoma).

3. Therapeutic Potential and Applications of Quercetin

Quercetin has been found to have therapeutic potential in various studies for the prevention and treatment various conditions as mentioned below.

3.1. Allergies, Asthma, Hay Fever and Hives

Some allergies, such as hay fever and hives, can benefit from the use of quercetin. It functions by stabilizing mast cell's membranes, which prevents the development and release of histamine and other allergic/inflammatory substances (Lombard, 2005; Thornhill and Kelly, 2000). Mast cells have been suggested as an immune gate to the brain, as well as environmental and emotional stress sensors, and are believed to be involved in neuropathologic processes e.g., multiple sclerosis. The protease tryptase, a mast cell product, may be related to neurodegenerative processes by activating specific receptors (PARs) in the brain, while interleukin (IL)-6 possibly induces neuro-degeneration and exacerbates dysfunction caused by other cytokines, or it may protect against the demyelinisation. Kempuraj et al. 2006, found that quercetin, a natural compound that inhibits mast cell secretion, induces a decrease in the release of tryptase and IL-6, as well as the downregulation of histidine decarboxylase (HDC) mRNA from human mast cell (HMC)-1. Since quercetin significantly inhibits mast cell tryptase, IL-6 release, and HDC mRNA transcription in the HMC-1 cell line, these findings suggest quercetin as a therapeutical compound for neurological diseases mediated by mast cell degranulation.

3.2. Antibacterial Activity

Quercetin appears to have antibacterial activity against nearly all bacteria strains recognized to cause respiratory, skin, gastrointestinal, and urinary disorders (Rigano et al., 2007; Adamczak et al., 2020).

3.3. Arthritis

Quercetin inhibits the activities of both cyclo-oxygenase and lipo-oxygenase, reducing the production of inflammatory mediators (Kim et al., 1998; Yoshimoto et al., 1983). Improvements in symptoms of peoples with rheumatoid arthritis (RA) have been reported after switching from a traditional western diet to a vegan diet rich in uncooked berries, vegetables and fruits which contain significant level of quercetin, among other antioxidants (Hanninen et al., 2000).The pathogenesis of RA has been linked to the abnormal formation of neutrophil extracellular traps (NETs). As a result, neutrophils are considered effective therapeutic targets in the treatment of RA. Previous research has shown that quercetin may be useful in the treatment of RA. According to Yuan et al. (2020) quercetin decreases the inflammation in RA mice by inhibiting neutrophil activity. Quercetin suppressed neutrophil infiltration and lowered inflammatory cytokine levels in the blood. Quercetin caused activated neutrophils to apoptose. In addition, quercetin suppressed autophagy, which prevented the development of NETs.

3.4. Cancers

The cancer has a multifactorial etiology which may include diet, genetics, and environment. But, it has been widely acknowledged that reactive oxygen and nitrogen species (ROS/RNS) play a critical role in the patho-physiological process. ROS/RON have been revealed to be carcinogenic and may cause their damaging effects by altering the gene expression (tumour suppressor gene, proto-oncogene) and by causing DNA damage, alter cell signalling pathways (ASK-1, MAPK, AP-1, PLA, NFkB). Plant-based diets have been shown to protect against different cancers in various studies including *in vitro* and *in vivo* laboratory research, clinical trials, and epidemiological investigations. In fact, it has been suggested that a diet rich in fruits and vegetables could prevent 7-31% of all cancers (Muhammad et al., 2006). Quercetin has been shown to prevent the growth of cancer cells in animal and test tube tests, including those from breast, colon, prostate, and lung cancers (Plakas et al., 1985). Quercetin's anti-oxidant properties protect DNA from reactive oxygen species-induced DNA damage, which can lead to mutational changes. A number of clinical studies have found an inverse relationship between the intake of quercetin and the subsequent occurrence of lung cancer (Knekt et al., 1997). It has been found that quercetin had an inhibitory effect on melanoma development, as well as influencing the invasive and metastatic potential in mice (Caltagirone et al., 2000). Quercetin, a bioflavonoid, may be a viable option for reducing cisplatin-induced nephrotoxicity (Heloisa et al., 2004). In addition, quercetin appears to prevent angiogenesis (Fotsis et al., 1997). In the human body, angiogenesis is usually a tightly regulated process. Cancers exhibit pathological, uncontrolled angiogenesis (Fan et al., 1995). Quercetin appears to be one of the most effective angiogenesis inhibitors (Paper, 1998). However, the mechanism behind flavonoids' anti-angiogenic activity is unknown. Protein kinase inhibition may be one potential mechanism (Oikawa et al., 1992).Since many of the PTKs are oncogenes, quercetin has the potential to be an important anti-cancer compound. Radiation-induced PKC activity was effectively blocked by quercetin. PKC activation is one of the ways a tumour cell can gain radio-resistance. Quercetin's suppression of PKC activity could be one way to avoid the development of radio resistance after radiotherapy (Varadkar et al., 2001). Chemical carcinogen e.g., hexavalent chromium [Cr (VI)] induced cell transformation is inhibited by quercetin (Pratheeshkumar et al., 2016; Han et al., 2016). Quercetin found to be promising both for the treatment and prevention of prostate cancer both in vitro as well as in vivo (Yang et al., 2015; Ward et al., 2018).

3.5. Coronary Heart Diseases

Consumption of the antioxidant quercetin protects against coronary heart disease (CHD), which is caused by oxidized LDL cholesterol (bad cholesterol). The daily consumption of flavonoids in food can lower the risk of CHD deaths in elderly men (Hertog et al., 1993; Hertog et al., 1995). In addition, a Japanese study found an inverse relationship between quercetin consumption and total plasma cholesterol levels (Arai et al., 2000). In dogs and monkeys, quercetin was found to be an important inhibitor of platelet aggregation (Osman et al., 1998). The inhibition of thromboxane A2 is the key antiplatelet aggregating effect (Tzeng et al., 1991). Quercetin prevents aortic smooth muscle cell proliferation and migration, as well as platelet aggregation and mitogen–activated protein kinase phosphorylation. These results add to the growing body of evidence supporting the use of quercetin in the prevention of cardiovascular disease (Hubbard et al., 2006).

3.6. Diabetes

Quercetin has been discovered to affect diabetic condition by many different ways e.g., by inhibiting the absorption of glucose in the intestine, by intervening the activities related to secretion and sensitization of insulin along with improved glucose utilization in peripheral tissues (Eid & Haddad, 2017).

3.7. Eye Disorders

Various eye diseases including cataracts and macular degeneration are affected by free radicals. Quercetin helps in preventing and treatment of these eye conditions by neutralizing free radicals. Moderate red wine intake (a source of quercetin) provided some protection against the development and progression of macular degeneration in a study of 3,072 adults with symptoms (Cai et al., 2000). Consumption of dark berries on a regular basis helps in preventing macular degeneration (Head et. al., 1999). McKay and Karamichos (2017) reviewed quercetin as a very promising flavonoid with substantial *in vitro* and *in vivo* beneficial effects for the ocular surface.

3.8. Gout

Elevated plasma uric acid concentration is a risk factor for gout. Because of its xanthine oxidase inhibitory properties, quercetin helps to alleviate gout symptoms by preventing the production of uric acid (Lio et al., 1986; Chang et al., 1993). Shi and Williamson in their double-blinded, randomised, placebo-controlled, cross-over trial found quercetin to decrease concentration of plasma uric acid (Shi and Williamson, 2016).

3.9. Neurodegenerative Disorders

According to Cornell University researchers, a strong antioxidant (quercetin) found in apples and vegetables appears to protect brain cells from oxidative stress, a tissue-damaging mechanism related to Alzheimer's disease (AD) and other neurodegenerative disorders (Heo et al., 2004). Quercetin appears to protect brain functions by preventing the development of fibrillated amyloid–beta, the senile plaque present in Alzheimer's brain (Tzeng et al., 1991). An experiment was carried out to show the effects of quercetin on cognitive performance in young and old, ethanol-intoxicated mice (animal model), with chronic quercetin treatment reversing cognitive deficits (Singh et al., 2003).

The anticatabolic activity of Ldopa plus carbidopa therapy may be improved by quercetin's ability to suppress COMT and MAO enzymes. The findings of this research strongly indicate quercetin may be a useful supplement to L-dopa therapy in Parkinson's disease (Singh et al., 2003). Quercetin has the ability to treat extrapyramidal side effects caused by neuroleptics, such as haloperidol (Naidu and Kulkarni, 2004). Quercetin as a strong antioxidant agent protects the brain cells from damage. Khan et al.(2020) have represented an extensive review on neuroprotective effects of quercetin in Alzheimer's disease. In a number of in vitro and in vivo models, quercetin has been found to have beneficial properties against general mechanisms of AD etiology. Quercetin protects the neural cells mainly by diminution of oxidative stress and neuro-inflammation, other anti-AD properties include inhibition of Aβ aggregation and tau phosphorylation. It restores acetylcholine levels by inhibiting acetylcholine hydrolysis. Low bioavailability due to extensive metabolization and low BBB penetrability limits the efficacy during the in-vivo studies. Quercetin and its analogues with greater gut and brain penetrability would likely to improve quercetin's clinical efficacy.

3.10. Osteoporosis

Quercetin is believed to increase bone mineral density, as in a study the bone density of elderly women who drank tea (a source of quercetin) was measured and compared with the bone mineral density of elderly women who did not drink tea. Hence the antioxidant quercetin in tea can help to prevent osteoporosis (Hegarty et al., 2000).

As per the results of study, (Derakhshanian et al., 2013) 150 mg quercetin/kg, alone or in combination with alendronate, can completely prevent Glucocorticoid-induced osteoporosis (GIO) by stimulating bone formation. Quercetin is thought to find therapeutic application in sustaining bone health as it enhances bone mass density (BMD), and bone formation, decreases osteocalcin levels and maintains bone biomechanics in ovariectomized rats (Yang et al., 2020).

3.11. Peptic Ulcer

Quercetin appears to be crucial in the prevention and treatment of peptic ulcer disease. It works as a gastroprotective agent by encouraging mucus secretion. *Helicobacter pylori*, an infectious bacterium, is thought to be the cause of many peptic ulcers. *In vitro* experiments have shown that quercetin inhibits the development of this bacterium (Martin et al., 1988; Alarcon-de-la-Lastra et al., 1994). Alkushi and Elsawy (2017) showed quercetin to protect gastric mucosa against indomethacin-induced gastric ulceration with its antioxidant effect.

3.12. Prostatitis

A double-blind placebo-controlled study was carried out and quercetin was found to be beneficial in category III chronic prostatitis (nonbacterial chronic prostatitis and prostodynia). For one month, thirty men with this condition were given either a placebo or 500 mg of quercetin twice daily. The National Institute of Health Chronic Prostatitis Score showed significant improvement in the treated population (Shoskes et al., 1999). Additional men were given the same amount of quercetin for one month in a follow-up unblind open trial, but this time quercetin was mixed with bromelain and papain, which could improve absorption. In this study, 82 percent of participants improved by at least 25%. Shoskes and Nickel (2011) discussed treatment of chronic prostatitis/chronic pelvic pain syndrome (CP/CPPS) with quercetin alone and as a part of multimodal therapy.

3.13. Viral Infections and COVID-19

In a research conducted by Wang et al. (1998) the antiviral activity of flavonoid was demonstrated. Herpes simplex virus, respiratory syncitial virus, and adenovirus are among the viruses that have been confirmed to be influenced by flavonoids. Quercetin has been shown to have anti-infective and anti-replicative properties against viruses.

Majority of experiments have been conducted in vitro, and only a little is known about flavonoids' antiviral effects in humans. A few substantiations are there which show that flavonoids in their glycon form tend to have a greater inhibitory effect on rotavirus infectivity than flavonoids in their aglycon form (Bae et al., 2000). Because of the worldwide spread of HIV, studies related to antiviral activity of flavonoids have largely been focused on HIV for last two three decades. Most of the studies have been focused on RNA guided DNA polymerase inhibitory activity or reverse transcriptase, though anti-integrase and anti-protease activities have also been reported in the *in vitro* studies (Vlietinck et al., 1998; Ng et al., 1997).

Wu et al., (2015) studied the inhibitory effect of quercetin on different strains of Influenza A viruses including A/Puerto Rico/8/34 (H1N1), A/FM-1/47/1 (H1N1), and A/Aichi/2/68 (H3N2) with an IC50 (half maximal inhibitory concentration) of 7.756 ± 1.097, 6.225 ± 0.467, and 2.738 ± 1.931 µg/mL respectively. Quercetin has been identified to interact with the HA2 subunit.

Due to overlapping antiviral and immunomodulatory properties and the ability of ascorbate to recycle quercetin, co-administration of vitamin C and quercetin has a synergistic antiviral impact (Colunga Biancatelli et al., 2020). In vitro and in silico studies confirmed that quercetin can interfere with different stages of the coronavirus entry and replication cycle e.g., PLpro, 3CLpro, and NTPase/helicase. Quercetin and its derivatives represent an important drug candidate compound due to lack of systemic toxicity and its pleiotropic activities (Agarwal et al., 2020; Di Pierro et al., 2020). Efficacy of quercetin supplementation for the treatment or prevention of COVID-19 has been tested with 3 three human clinical trials, it was found that though beneficial effect on the incidence and duration of respiratory tract infections in certain populations have been observed with oral quercetin, but further research is required to recommend quercetin supplementation as a therapy for the treatment or prevention of COVID-19 (Aucoin et al., 2020). A case reported by Moskowitz et al., (2020) an 18-year-old asthmatic male with symptoms of COVID19, responded strongly to quercetin.SARS-CoV-2 is a betacoronavirus, the causal agent of COVID-19, excessive inflammation is considered one of the main life-threatening conditions in patients. Quercetin, as an anti-inflammatory, antioxidant compound, is probably a potential treatment for severe inflammation in patients with COVID-19 (Saeedi-Boroujeni and Mahmoudian-Sani, 2021).

Conclusion

Quercetin has been shown to have beneficial medicinal properties in the treatment of a variety of diseases. Quercetin has potential to be a new medication that can prevent and treat a number of diseases. Its potent antioxidant, anti-inflammatory, and anti-tumor properties hold great promise for clinical use. Quercetin is a natural antioxidant that can be fed to animals. As quercetin performs antioxidant functions in the body, it can also enhance physical functions and reduce stress responses. Quercetin's broad-spectrum antimicrobial properties may be used to prevent and treat a variety of infectious bacterial diseases, as well as provide treatment options that minimize antibiotic usage, which has significant consequences. Quercetin has been found one of the promising flavonoids in the various studies carried out by various researchers time to time. Quercetin has the potential to be used for management of various diseases as safe, effective, and alternative medicine or combination based medicine. Most of the studies have been carried out *in-vitro* or using animal models, therefore, extensive research on various aspects (including dosage, combinations, clinical trials, mechanism of action and others) are required to establish the quercetin as therapeutic molecule and to explore the quercetin based prevention and management of various diseases.

References

Adamczak, A., Ozarowski, M., & Karpinski, T. M. (2019). Antibacterial Activity of Some Flavonoids and Organic Acids Widely Distributed in Plants. *Journal of clinical medicine*, *9*(1), 109. https://doi.org/10.3390/jcm9010109.

Afanasev, I. B., Dorozhko, A. I., Brodskii, A. V., Kostyuk, V. A., & Potapovitch, A. I. (1989). Chelating and free radical scavenging mechanisms of inhibitory action of rutin and quercetin in lipid peroxidation. *Biochemical pharmacology*, 38(11), 1763–1769. https://doi.org/10.1016/0006-2952(89)90410-3.

Aggarwal B. B. (2000). Tumour necrosis factors receptor associated signalling molecules and their role in activation of apoptosis, JNK and NF-kappaB. *Annals of the rheumatic diseases*, 59 Suppl 1(Suppl 1), i6–i16. https://doi.org/10.1136/ard.59.suppl_1.i6.

Aggarwal, B. B., Samanta, A., & Feldmann, M. (2001). TNF-α, p. 413. In: *Cytokine reference*, Oppenheim, J. J., Feldman, M., Durum, S. K., Hirano, T., Vilcek, J. & Nicola, N. A. (Ed.), ISBN: 978-012-25-2670-1.

Agrawal, P. K., Agrawal, C., & Blunden, G. (2020). Quercetin: Antiviral Significance and Possible COVID-19 Integrative Considerations. *Natural Product Communications*. https://doi.org/10.1177/1934578X20976293.

Alarcon de la Lastra, C., Martin, M. J., & Motilva, V. (1994). Antiulcer and gastroprotective effects of quercetin: a gross and histologic study. *Pharmacology*. 48(1), 56–62. https://doi.org/10.1159/000139162.

Alkushi, A., & Elsawy, N. (2017). Quercetin attenuates, indomethacin-induced acute gastric ulcer in rats. *Folia morphologica*, *76*(2), 252–261. https://doi.org/10.5603/FM.a2016.0067.

Anand David, A. V., Arulmoli, R., & Parasuraman, S. (2016). Overviews of Biological Importance of Quercetin: A Bioactive Flavonoid. *Pharmacognosy reviews*, *10*(20), 84–89. https://doi.org/10.4103/0973-7847.194044.

Arai, Y., Watanabe, S., Kimira, M., Shimoi, K., Mochizuki, R., & Kinae, N. (2000). Dietary intakes of flavonols, flavones and isoflavones by Japanese women and the inverse correlation between quercetin intake and plasma LDL cholesterol concentration. *The Journal of nutrition*, *130*(9), 2243–2250. https://doi.org/10.1093/jn/130.9.2243.

Ares, J. J., & Outt, P. E. (1998). Gastroprotective agents for the prevention of NSAID-induced gastropathy. *Current pharmaceutical design*, *4*(1), 17–36.

Aucoin, M., Cooley, K., Saunders, P. R., Cardozo, V., Remy, D., Cramer, H., Neyre Abad, C., & Hannan, N. (2020). The effect of quercetin on the prevention or treatment of COVID-19 and other respiratory tract infections in humans: A rapid review. *Advances in integrative medicine*, 7(4), 247–251. https://doi.org/10.1016/j.aimed.2020.07.007.

Bae, E. A., Han, M. J., Lee, M., & Kim, D. H. (2000). In vitro inhibitory effect of some flavonoids on rotavirus infectivity. *Biological & pharmaceutical bulletin*, *23*(9), 1122–1124. https://doi.org/10.1248/bpb.23.1122.

Baghel, S. S., Shrivastava, R. S., Agrawal, P., & Rajput, S. (2012). A review of Quercetin: Antioxidant and Anticancer properties. *World Journal of Pharmacy and Pharmaceutical Sciences*, 1(1), 146-160.

Bahorun, T., Soobrattee, M. A., Luximon-Ramma, V., & Aruoma, O. I. (2006). Free Radicals and Antioxidants in Cardiovascular Health and Disease. *Internet Journal of Medical Update*, 1(2), 24-40.

Baier, W. E. (1955). Chemistry of bioflavonoids. *Annals of the New York Academy of Sciences*, *61*(3), 639–645. https://doi.org/10.1111/j.1749-6632.1955.tb42516.x.

Batiha, G. E., Beshbishy, A. M., Ikram, M., Mulla, Z. S., El-Hack, M., Taha, A. E., Algammal, A. M., & Elewa, Y. (2020). The Pharmacological Activity, Biochemical Properties, and Pharmacokinetics of the Major Natural Polyphenolic Flavonoid: Quercetin. *Foods (Basel, Switzerland)*, *9*(3), 374. https://doi.org/10.3390/foods9030374.

Bennett, J. P., Gomperts, B. D., & Wollenweber, E. (1981). Inhibitory effects of natural flavonoids on secretion from mast cells and neutrophils. *Arzneimittel-Forschung*, *31*(3), 433–437.

Bhagwat, S., Haytowitz, D. B., Holden, J. M. (Ret.) (2013). USDA Database for the Flavonoid Content of Selected Foods, Release 3.1. U.S. Department of Agriculture, Agricultural Research Service. *Nutrient Data.* https://www.ars.usda.gov/ARSUserFiles/80400525/Data/Flav/Flav3-1.pdf [Accessed on March 13, 2021].

Busse, W. W., Kopp, D. E., & Middleton, E., Jr (1984). Flavonoid modulation of human neutrophil function. *The Journal of allergy and clinical immunology*, *73*(6), 801–809. https://doi.org/10.1016/0091-6749(84)90450-0.

Cai, J., Nelson, K. C., Wu, M., Sternberg, P., Jr, & Jones, D. P. (2000). Oxidative damage and protection of the RPE. *Progress in retinal and eye research*, *19*(2), 205–221. https://doi.org/10.1016/s1350-9462(99)00009-9.

Calamia K. T. (2003). Current and future use of anti-TNF agents in the treatment of autoimmune, inflammatory disorders. *Advances in experimental medicine and biology*, *528*, 545–549. https://doi.org/10.1007/0-306-48382-3_110.

Caltagirone, S., Rossi, C., Poggi, A., Ranelletti, F. O., Natali, P. G., Brunetti, M., Aiello, F. B., & Piantelli, M. (2000). Flavonoids apigenin and quercetin inhibit melanoma growth and metastatic potential. *International journal of cancer*, *87*(4), 595–600. https://doi.org/10.1002/1097-0215(20000815)87:4<595::aid-ijc21>3.0.co;2-5.

Chang, W. S., Lee, Y. J., Lu, F. J., & Chiang, H. C. (1993). Inhibitory effects of flavonoids on xanthine oxidase. *Anticancer research*, *13*(6A), 2165–2170.

Colunga Biancatelli, R., Berrill, M., Catravas, J. D., & Marik, P. E. (2020). Quercetin and Vitamin C: An Experimental, Synergistic Therapy for the Prevention and Treatment of SARS-CoV-2 Related Disease (COVID-19). *Frontiers in immunology*, *11*, 1451. https://doi.org/10.3389/fimmu.2020.01451.

de Groot H. (1994). Reactive oxygen species in tissue injury. *Hepato-gastroenterology*, 41(4), 328–332.

Della Loggia, R., Ragazzi, E., Tubaro, A., Fassina, G., & Vertua, R. (1988). Anti-inflammatory activity of benzopyrones that are inhibitors of cyclo- and lipo-oxygenase. *Pharmacological research communications*, 20 Suppl 5, 91–94. https://doi.org/10.1016/s0031-6989(88)80849-x.

Derakhshanian, H., Djalali, M., Djazayery, A., Nourijelyani, K., Ghadbeigi, S., Pishva, H., Saedisomeolia, A., Bahremand, A., & Dehpour, A. R. (2013). Quercetin prevents experimental glucocorticoid-induced osteoporosis: a comparative study with alendronate. *Canadian journal of physiology and pharmacology*, 91(5), 380–385. https://doi.org/10.1139/cjpp-2012-0190.

Di Pierro, F., Khan, A., Bertuccioli, A., Maffioli, P., Derosa, G., Khan, S., Khan, B. A., Nigar, R., Ujjan, I., & Devrajani, B. R. (2020). Quercetin Phytosome® as a potential candidate for managing COVID-19. *Minerva gastroenterologica e dietologica*, 10.23736/S1121-421X.20.02771-3. Advance online publication. https://doi.org/10.23736/S1121-421X.20.02771-3.

Eid, H. M., & Haddad, P. S. (2017). The Antidiabetic Potential of Quercetin: Underlying Mechanisms. *Current medicinal chemistry*, 24(4), 355–364. https://doi.org/10.2174/0929867323666160909153707.

Fan, T. P., Jaggar, R., & Bicknell, R. (1995). Controlling the vasculature: angiogenesis, anti-angiogenesis and vascular targeting of gene therapy. *Trends in pharmacological sciences,* 16(2), 57–66. https://doi.org/10.1016/s0165-6147(00)88979-8.

Ferrali, M., Signorini, C., Caciotti, B., Sugherini, L., Ciccoli, L., Giachetti, D., & Comporti, M. (1997). Protection against oxidative damage of erythrocyte membrane by the flavonoid quercetin and its relation to iron chelating activity. *FEBS letters*, 416(2), 123–129. https://doi.org/10.1016/s0014-5793(97)01182-4.

Fischer, C., Speth, V., Fleig-Eberenz, S., & Neuhaus, G. (1997). Induction of Zygotic Polyembryos in Wheat: Influence of Auxin Polar Transport. *The Plant cell*, 9(10), 1767–1780. https://doi.org/10.1105/tpc.9.10.1767.

Fotsis, T., Pepper, M. S., Aktas, E., Breit, S., Rasku, S., Adlercreutz, H., Wähälä, K., Montesano, R., & Schweigerer, L. (1997). Flavonoids, dietary-derived inhibitors of cell proliferation and in vitro angiogenesis. *Cancer research*, 57(14), 2916–2921.

Fox, C. C., Wolf, E. J., Kagey-Sobotka, A., & Lichtenstein, L. M. (1988). Comparison of human lung and intestinal mast cells. *The Journal of allergy and clinical immunology,* 81(1), 89–94. https://doi.org/10.1016/0091-6749(88)90225-4.

Fraga, C. G., Martino, V. S., Ferraro, G. E., Coussio, J. D., & Boveris, A. (1987). Flavonoids as antioxidants evaluated by in vitro and in situ liver chemiluminescence. *Biochemical pharmacology,* 36(5), 717–720. https://doi.org/10.1016/0006-2952(87) 90724-6.

Francescato, H. D., Coimbra, T. M., Costa, R. S., & Bianchi, M. (2004). Protective effect of quercetin on the evolution of cisplatin-induced acute tubular necrosis. *Kidney & blood pressure research*, *27*(3), 148–158. https://doi.org/10.1159/000078309.

Friesenecker, B., Tsai, A. G., Allegra, C., & Intaglietta, M. (1994). Oral administration of purified micronized flavonoid fraction suppresses leukocyte adhesion in ischemia-reperfusion injury: in vivo observations in the hamster skin fold. *International journal of microcirculation, clinical and experimental,* 14(1-2), 50–55. https://doi.org/10.1159/000178206.

Grace P. A. (1994). Ischaemia-reperfusion injury. *The British journal of surgery, 81*(5), 637–647. https://doi.org/10.1002/bjs.1800810504.

Halliwell B. (1994). Free radicals, antioxidants, and human disease: curiosity, cause, or consequence? *Lancet (London, England), 344*(8924), 721–724. https://doi.org/10.1016/s0140-6736(94)92211-x.

Halliwell B. (1995). How to characterize an antioxidant: an update. *Biochemical Society symposium*, *61*, 73–101. https://doi.org/10.1042/bss0610073.

Han, M., Song, Y., & Zhang, X. (2016). Quercetin Suppresses the Migration and Invasion in Human Colon Cancer Caco-2 Cells Through Regulating Toll-like Receptor 4/Nuclear Factor-kappa B Pathway. *Pharmacognosy magazine*, *12*(Suppl 2), S237–S244. https://doi.org/10.4103/0973-1296.182154.

Hänninen, Kaartinen, K., Rauma, A. L., Nenonen, M., Törrönen, R., Häkkinen, A. S., Adlercreutz, H., & Laakso, J. (2000). Antioxidants in vegan diet and rheumatic disorders. *Toxicology*, *155*(1-3), 45–53. https://doi.org/10.1016/s0300-483x(00) 00276-6.

Havsteen B. (1983). Flavonoids, a class of natural products of high pharmacological potency. *Biochemical pharmacology*, *32*(7), 1141–1148. https://doi.org/10.1016/0006-2952(83)90262-9.

Head K. A. (1999). Natural therapies for ocular disorders, part one: diseases of the retina. *Alternative medicine review : a journal of clinical therapeutic*, *4*(5), 342–359.

Hegarty, V. M., May, H. M., & Khaw, K. T. (2000). Tea drinking and bone mineral density in older women. *The American journal of clinical nutrition*, *71*(4), 1003–1007. https://doi.org/10.1093/ajcn/71.4.1003.

Heo, H. J., Kim, D. O., Choi, S. J., Shin, D. H., Lee, C. Y. (2004) Apple Phenolics Protect in Vitro Oxidative Stress-induced Neuronal Cell Death. *Journal of Food Science, 69*(9), S357-S360. https://doi.org/10.1111/j.1365-2621.2004.tb09949.x.

Hertog, M. G., Feskens, E. J., Hollman, P. C., Katan, M. B., & Kromhout, D. (1993). Dietary antioxidant flavonoids and risk of coronary heart disease: the Zutphen Elderly Study. *Lancet (London, England)*, *342*(8878), 1007–1011. https://doi.org/10.1016/0140-6736(93)92876-u.

Hertog, M. G., Kromhout, D., Aravanis, C., Blackburn, H., Buzina, R., Fidanza, F., Giampaoli, S., Jansen, A., Menotti, A., & Nedeljkovic, S. (1995). Flavonoid intake and long-term risk of coronary heart disease and cancer in the seven countries study. *Archives of internal medicine*, *155*(4), 381–386.

Hollman, P. C., Bijsman, M. N., van Gameren, Y., Cnossen, E. P., de Vries, J. H., & Katan, M. B. (1999). The sugar moiety is a major determinant of the absorption of dietary flavonoid glycosides in man. *Free radical research*, *31*(6), 569–573. https://doi.org/10.1080/10715769900301141.

Hubbard, G. P., Wolffram, S., de Vos, R., Bovy, A., Gibbins, J. M., & Lovegrove, J. A. (2006). Ingestion of onion soup high in quercetin inhibits platelet aggregation and essential components of the collagen-stimulated platelet activation pathway in man: a pilot study. *The British journal of nutrition*, *96*(3), 482–488.

Huk, I., Brovkovych, V., Nanobash Vili, J., Weigel, G., Neumayer, C., Partyka, L., Patton, S., & Malinski, T. (1998). Bioflavonoid quercetin scavenges superoxide and increases nitric oxide concentration in ischaemia-reperfusion injury: an experimental study. *The British journal of surgery*, *85*(8), 1080–1085. https://doi.org/10.1046/j.1365-2168.1998.00787.x.

Iio, M., Ono, Y., Kai, S., & Fukumoto, M. (1986). Effects of flavonoids on xanthine oxidation as well as on cytochrome c reduction by milk xanthine oxidase. *Journal of nutritional science and vitaminology*, *32*(6), 635–642. https://doi.org/10.3177/jnsv.32.635.

Kaneuchi, M., Sasaki, M., Tanaka, Y., Sakuragi, N., Fujimoto, S., & Dahiya, R. (2003). Quercetin regulates growth of Ishikawa cells through the suppression of EGF and cyclin D1. *International journal of oncology*, *22*(1), 159–164.

Kempuraj, D., Castellani, M. L., Petrarca, C., Frydas, S., Conti, P., Theoharides, T. C., & Vecchiet, J. (2006). Inhibitory effect of quercetin on tryptase and interleukin-6 release, and histidine decarboxylase mRNA transcription by human mast cell-1 cell line. *Clinical and experimental medicine*, *6*(4), 150–156. https://doi.org/10.1007/s10238-006-0114-7.

Kerry, N. L., & Abbey, M. (1997). Red wine and fractionated phenolic compounds prepared from red wine inhibit low density lipoprotein oxidation in vitro. *Atherosclerosis*, *135*(1), 93–102. https://doi.org/10.1016/s0021-9150(97)00156-1.

Khan, H., Ullah, H., Aschner, M., Cheang, W. S., & Akkol, E. K. (2019). Neuroprotective Effects of Quercetin in Alzheimer's Disease. *Biomolecules*, *10*(1), 59. https://doi.org/10.3390/biom10010059.

Kim, H. P., Mani, I., Iversen, L., & Ziboh, V. A. (1998). Effects of naturally-occurring flavonoids and biflavonoids on epidermal cyclooxygenase and lipoxygenase from guinea-pigs. *Prostaglandins, leukotrienes, and essential fatty acids*, *58*(1), 17–24. https://doi.org/10.1016/s0952-3278(98)90125-9.

Knekt, P., Järvinen, R., Seppänen, R., Hellövaara, M., Teppo, L., Pukkala, E., & Aromaa, A. (1997). Dietary flavonoids and the risk of lung cancer and other malignant neoplasms. *American journal of epidemiology*, *146*(3), 223–230. https://doi.org/10.1093/oxfordjournals.aje.a009257.

Lakhanpal, P., Rai, D. K. (2007) Quercetin: A Versatile Flavonoid. *Internet Journal of Medical Update* 2(2), 22-37.

Lombard, K. A. (2005). Investigation of the flavonol quercetin in onion (Allium cepa L.) by highperformance liquid chromatography (HPLC) and spectrophotometric methodology. M.S. Thesis. *Journal of Food Composition and Analysis18*, 635-645.

Mandel, S., Weinreb, O., Amit, T., & Youdim, M. B. (2004). Cell signaling pathways in the neuroprotective actions of the green tea polyphenol (-)-epigallocatechin-3-gallate: implications for neurodegenerative diseases. *Journal of neurochemistry*, *88*(6), 1555–1569. https://doi.org/10.1046/j.1471-4159.2003.02291.x.

Mariani, C., Braca, A., Vitalini, S., De Tommasi, N., Visioli, F., & Fico, G. (2008). Flavonoid characterization and in vitro antioxidant activity of Aconitum anthora L. (Ranunculaceae). *Phytochemistry*, *69*(5), 1220–1226. https://doi.org/10.1016/j.phytochem.2007.12.009.

Martín, M. J., La-Casa, C., Alarcón-de-la-Lastra, C., Cabeza, J., Villegas, I., & Motilva, V. (1998). Anti-oxidant mechanisms involved in gastroprotective effects of quercetin. *Zeitschrift fur Naturforschung. C, Journal of biosciences*, *53*(1-2), 82–88. https://doi.org/10.1515/znc-1998-1-215.

McKay, T. B., & Karamichos, D. (2017). Quercetin and the ocular surface: What we know and where we are going. *Experimental biology and medicine (Maywood, N.J.)*, *242*(6), 565–572. https://doi.org/10.1177/1535370216685187.

Middleton, E., Jr, & Anne, S. (1995). Quercetin inhibits lipopolysaccharide-induced expression of endothelial cell intracellular adhesion molecule-1. *International archives of allergy and immunology*, *107*(1-3), 435–436. https://doi.org/10.1159/000237071.

Middleton, E., Jr, & Drzewiecki, G. (1985). Naturally occurring flavonoids and human basophil histamine release. *International archives of allergy and applied immunology*, *77*(1-2), 155–157. https://doi.org/10.1159/000233771.

Middleton, E., Jr, Drzewiecki, G., & Krishnarao, D. (1981). Quercetin: an inhibitor of antigen-induced human basophil histamine release. *Journal of immunology (Baltimore, Md.: 1950)*, *127*(2), 546–550.

Miean, K. H., & Mohamed, S. (2001). Flavonoid (myricetin, quercetin, kaempferol, luteolin, and apigenin) content of edible tropical plants. *Journal of agricultural and food chemistry*, *49*(6), 3106–3112. https://doi.org/10.1021/jf000892m.

Mohle, B., Heller, W., Wellmann, E. (1985) UV-Induced biosynthesis of quercetin 3-O-β-D-glucuronide in dill cell cultures. *Phytochemistry, 24*(3), 465-467. https://doi.org/10.1016/S0031-9422(00)80748-7.

Moskowitz, D. W., Sanchez-Gonzalez, M. A., Marinelli, E. R., & Day, K. (2020) Quercetin for COVID19? *Diabetes Complications4*(2), 1-2. (ISSN 2639-9326).

Naidu, P. S., & Kulkarni, S. K. (2004). Quercetin, a bioflavonoid, reverses haloperidol-induced catalepsy. *Methods and findings in experimental and clinical pharmacology*, *26*(5), 323–326. https://doi.org/10.1358/mf.2004.26.5.831321.

Ng, T. B., Huang, B., Fong, W. P., & Yeung, H. W. (1997). Anti-human immunodeficiency virus (anti-HIV) natural products with special emphasis on HIV reverse transcriptase inhibitors. *Life sciences*, *61*(10), 933–949. https://doi.org/10.1016/s0024-3205(97)00245-2.

Nishimuro, H., Ohnishi, H., Sato, M., Ohnishi-Kameyama, M., Matsunaga, I., Naito, S., Ippoushi, K., Oike, H., Nagata, T., Akasaka, H., Saitoh, S., Shimamoto, K., & Kobori, M. (2015). Estimated daily intake and seasonal food sources of quercetin in Japan. *Nutrients*, *7*(4), 2345–2358. https://doi.org/10.3390/nu7042345.

Nishino, H., Naitoh, E., Iwashima, A., & Umezawa, K. (1984). Quercetin interacts with calmodulin, a calcium regulatory protein. *Experientia*, *40*(2), 184–185. https://doi.org/10.1007/BF01963590.

Oikawa, T., Shimamura, M., Ashino, H., Nakamura, O., Kanayasu, T., Morita, I., & Murota, S. (1992). Inhibition of angiogenesis by staurosporine, a potent protein kinase inhibitor. *The Journal of antibiotics*, *45*(7), 1155–1160. https://doi.org/10.7164/antibiotics.45.1155.

Osman, H. E., Maalej, N., Shanmuganayagam, D., & Folts, J. D. (1998). Grape juice but not orange or grapefruit juice inhibits platelet activity in dogs and monkeys. *The Journal of nutrition*, *128*(12), 2307–2312. https://doi.org/10.1093/jn/128.12.2307.

Otsuka, H., Inaba, M., Fujikura, T., & Kunitomo, M. (1995). Histochemical and functional characteristics of metachromatic cells in the nasal epithelium in allergic rhinitis: studies of nasal scrapings and their dispersed cells. *The Journal of allergy and clinical immunology*, *96*(4), 528–536. https://doi.org/10.1016/s0091-6749(95)70297-0.

Paper D. H. (1998). Natural products as angiogenesis inhibitors. *Planta medica, 64*(8), 686–695. https://doi.org/10.1055/s-2006-957559.

Patil, B. S., Pike, L. M. (1995). Distribution of quercetin content in different rings of various colored onion (*Allium cepa* L.) cultivars. *Journal of Horticultural Sciences, 70*(4), 643-650. https://doi.org/10.1080/14620316.1995.11515338.

Pearce, F. L., Befus, A. D., & Bienenstock, J. (1984). Mucosal mast cells. III. Effect of quercetin and other flavonoids on antigen-induced histamine secretion from rat intestinal mast cells. *The Journal of allergy and clinical immunology*, *73*(6), 819–823. https://doi.org/10.1016/0091-6749(84)90453-6.

Plakas, S. M., Lee, T. C., & Wolke, R. E. (1985). Absence of overt toxicity from feeding the flavonol, quercetin, to rainbow trout (Salmo gairdneri). *Food and chemical toxicology: an international journal published for the British Industrial Biological*

Research Association, *23*(12), 1077–1080. https://doi.org/10.1016/0278-6915(85)90055-9.

Pratheeshkumar, P., Son, Y. O., Divya, S. P., Wang, L., Turcios, L., Roy, R. V., Hitron, J. A., Kim, D., Dai, J., Asha, P., Zhang, Z., & Shi, X. (2016). Quercetin inhibits Cr(VI)-induced malignant cell transformation by targeting miR-21-PDCD4 signaling pathway. *Oncotarget*, *8*(32), 52118–52131. https://doi.org/10.18632/oncotarget.10130.

Rigano, D., Formisano, C., Basile, A., Lavitola, A., Senatore, F., Rosselli, S., & Bruno, M. (2007). Antibacterial activity of flavonoids and phenylpropanoids from Marrubium globosum ssp. libanoticum. *Phytotherapy research: PTR*, *21*(4), 395–397. https://doi.org/10.1002/ptr.2061.

Ross, J. A., & Kasum, C. M. (2002). Dietary flavonoids: bioavailability, metabolic effects, and safety. *Annual review of nutrition*, *22*, 19–34. https://doi.org/10.1146/annurev.nutr.22.111401.144957.

Saeedi-Boroujeni, A., Mahmoudian-Sani, M. R. (2021). Anti-inflammatory potential of Quercetin in COVID-19 treatment. *Journal of Inflammation,18*, 3 https://doi.org/10.1186/s12950-021-00268-6.

Sampson, L., Rimm, E., Hollman, P. C., de Vries, J. H., & Katan, M. B. (2002). Flavonol and flavone intakes in US health professionals. *Journal of the American Dietetic Association*, *102*(10), 1414–1420. https://doi.org/10.1016/s0002-8223(02)90314-7.

Sanhueza, J., Valdes, J., Campos, R., Garrido, A., & Valenzuela, A. (1992). Changes in the xanthine dehydrogenase/xanthine oxidase ratio in the rat kidney subjected to ischemia-reperfusion stress: preventive effect of some flavonoids. *Research communications in chemical pathology and pharmacology*, *78*(2), 211–218.

Santos, A. C., Uyemura, S. A., Lopes, J. L., Bazon, J. N., Mingatto, F. E., & Curti, C. (1998). Effect of naturally occurring flavonoids on lipid peroxidation and membrane permeability transition in mitochondria. *Free radical biology & medicine*, *24*(9), 1455–1461. https://doi.org/10.1016/s0891-5849(98)00003-3.

Shi, Y., & Williamson, G. (2016). Quercetin lowers plasma uric acid in pre-hyperuricaemic males: a randomised, double-blinded, placebo-controlled, cross-over trial. *The British journal of nutrition*, *115*(5), 800–806. https://doi.org/10.1017/S0007114515005310.

Shoskes D. A. (1998). Effect of bioflavonoids quercetin and curcumin on ischemic renal injury: a new class of renoprotective agents. *Transplantation*, *66*(2), 147–152. https://doi.org/10.1097/00007890-199807270-00001.

Shoskes, D. A., & Nickel, J. C. (2011). Quercetin for chronic prostatitis/chronic pelvic pain syndrome. *The Urologic clinics of North America*, *38*(3), 279–284. https://doi.org/10.1016/j.ucl.2011.05.003.

Shoskes, D. A., Zeitlin, S. I., Shahed, A., & Rajfer, J. (1999). Quercetin in men with category III chronic prostatitis: a preliminary prospective, double-blind, placebo-controlled trial. *Urology*, *54*(6), 960–963. https://doi.org/10.1016/s0090-4295(99)00358-1.

Shutenko, Z., Henry, Y., Pinard, E., Seylaz, J., Potier, P., Berthet, F., Girard, P., & Sercombe, R. (1999). Influence of the antioxidant quercetin in vivo on the level of nitric oxide determined by electron paramagnetic resonance in rat brain during global ischemia and reperfusion. *Biochemical pharmacology*, *57*(2), 199–208. https://doi.org/10.1016/s0006-2952(98)00296-2.

Singh, A., Naidu, P. S., & Kulkarni, S. K. (2003). Quercetin potentiates L-Dopa reversal of drug-induced catalepsy in rats: possible COMT/MAO inhibition. *Pharmacology*, *68*(2), 81–88. https://doi.org/10.1159/000069533.

Singh, A., Naidu, P. S., & Kulkarni, S. K. (2003). Reversal of aging and chronic ethanol-induced cognitive dysfunction by quercetin a bioflavonoid. *Free radical research*, *37*(11), 1245–1252. https://doi.org/10.1080/10715760310001616014.

Soobrattee, M. A., Bahorun, T., & Aruoma, O. I. (2006). Chemopreventive actions of polyphenolic compounds in cancer. *BioFactors (Oxford, England)*, *27*(1-4), 19–35. https://doi.org/10.1002/biof.5520270103.

Sorata, Y., Takahama, U., & Kimura, M. (1984). Protective effect of quercetin and rutin on photosensitized lysis of human erythrocytes in the presence of hematoporphyrin. *Biochimica et biophysica acta*, *799*(3), 313–317. https://doi.org/10.1016/0304-4165(84)90276-9.

Sun, C., Wang, H., Wang, D., Chen, Y., Zhao, Y., & Xia, W. (2015). Using an FFQ to assess intakes of dietary flavonols and flavones among female adolescents in the Suihua area of northern China. *Public health nutrition*, *18*(4), 632–639. https://doi.org/10.1017/S1368980014000780.

Taylor, P. C., Williams, R. O., & Feldmann, M. (2004). Tumour necrosis factor alpha as a therapeutic target for immune-mediated inflammatory diseases. *Current opinion in biotechnology*, *15*(6), 557–563. https://doi.org/10.1016/j.copbio.2004.09.005.

Thornhill, S. M., & Kelly, A. M. (2000). Natural treatment of perennial allergic rhinitis. *Alternative medicine review: a journal of clinical therapeutic*, *5*(5), 448–454.

Tzeng, S. H., Ko, W. C., Ko, F. N., & Teng, C. M. (1991). Inhibition of platelet aggregation by some flavonoids. *Thrombosis research*, *64*(1), 91–100. https://doi.org/10.1016/0049-3848(91)90208-e.

van Acker, S. A., Tromp, M. N., Haenen, G. R., van der Vijgh, W. J., & Bast, A. (1995). Flavonoids as scavengers of nitric oxide radical. *Biochemical and biophysical research communications*, *214*(3), 755–759. https://doi.org/10.1006/bbrc.1995.2350.

Varadkar, P., Dubey, P., Krishna, M., & Verma, N. (2001). Modulation of radiation-induced protein kinase C activity by phenolics. *Journal of radiological protection: official journal of the Society for Radiological Protection*, *21*(4), 361–370. https://doi.org/10.1088/0952-4746/21/4/304.

Vlietinck, A. J., De Bruyne, T., Apers, S., & Pieters, L. A. (1998). Plant-derived leading compounds for chemotherapy of human immunodeficiency virus (HIV) infection. *Planta medica*, *64*(2), 97–109. https://doi.org/10.1055/s-2006-957384.

Wajant, H., Henkler, F., & Scheurich, P. (2001). The TNF-receptor-associated factor family: scaffold molecules for cytokine receptors, kinases and their regulators. *Cellular signalling*, *13*(6), 389–400. https://doi.org/10.1016/s0898-6568(01)00160-7.

Wang, H. K., Xia, Y., Yang, Z. Y., Natschke, S. L., & Lee, K. H. (1998). Recent advances in the discovery and development of flavonoids and their analogues as antitumor and anti-HIV agents. *Advances in experimental medicine and biology*, *439*, 191–225. https://doi.org/10.1007/978-1-4615-5335-9_15.

Ward, A. B., Mir, H., Kapur, N., Gales, D. N., Carriere, P. P., & Singh, S. (2018). Quercetin inhibits prostate cancer by attenuating cell survival and inhibiting anti-apoptotic pathways. *World journal of surgical oncology*, *16*(1), 108. https://doi.org/10.1186/s12957-018-1400-z.

Wegener, T., Fintelmann, V. (1999). Flavonoids and Bioactivity. *Wein Med Wochem Schr*, 149, 241-247.

Wu, W., Li, R., Li, X., He, J., Jiang, S., Liu, S., & Yang, J. (2015). Quercetin as an Antiviral Agent Inhibits Influenza A Virus (IAV) Entry. *Viruses*, *8*(1), 6. https://doi.org/10.3390/v8010006.

Yang, F., Song, L., Wang, H., Wang, J., Xu, Z., & Xing, N. (2015). Quercetin in prostate cancer: Chemotherapeutic and chemopreventive effects, mechanisms and clinical application potential (Review). *Oncology reports*, *33*(6), 2659–2668. https://doi.org/10.3892/or.2015.3886.

Yang, H., Liu, J., Wang, M., Wang, L., Zhang, L., & Zhang, F. (2020). Effect of quercetin on bone metabolism and serum osteocalcin in osteoporotic rats. *Tropical Journal of Pharmaceutical Research, 19* (2), 277-281. http://dx.doi.org/10.4314/tjpr.v19i2.9.

Yoshimoto, T., Furukawa, M., Yamamoto, S., Horie, T., & Watanabe-Kohno, S. (1983). Flavonoids: potent inhibitors of arachidonate 5-lipoxygenase. *Biochemical and biophysical research communications*, *116*(2), 612–618. https://doi.org/10.1016/0006-291x(83)90568-5.

Yuan, K., Zhu, Q., Lu, Q., Jiang, H., Zhu, M., Li, X., Huang, G. & Xu, A. (2020). Quercetin alleviates rheumatoid arthritis by inhibiting neutrophil inflammatory activities. *The Journal of Nutritional Biochemistry 84*, https://doi.org/10.1016/j.jnutbio.2020.108454.

Zamora-Ros, R., Andres-Lacueva, C., Lamuela-Raventós, R. M., Berenguer, T., Jakszyn, P., Barricarte, A., Ardanaz, E., Amiano, P., Dorronsoro, M., Larrañaga, N., Martínez, C., Sánchez, M. J., Navarro, C., Chirlaque, M. D., Tormo, M. J., Quirós, J. R., & González, C. A. (2010). Estimation of dietary sources and flavonoid intake in a Spanish adult population (EPIC-Spain). *Journal of the American Dietetic Association*, *110*(3), 390–398. https://doi.org/10.1016/j.jada.2009.11.024.

Zhang, Y., Li, Y., Cao, C., Cao, J., Chen, W., Zhang, Y., Wang, C., Wang, J., Zhang, X., & Zhao, X. (2010). Dietary flavonol and flavone intakes and their major food sources in Chinese adults. *Nutrition and cancer,* 62(8), 1120–1127. https://doi.org/10.1080/01635581.2010.513800.

Chapter 3

Biological Activities and Therapeutic Potential of the Citrus Flavonoid Naringin

Rishabh Srivastava[1], Manish Kumar[2,*], Diwakar Aggarwal[3] and Mukesh Yadav[3]

[1]Pontika Aerotech Pvt. Ltd., Paonta Sahib, Himanchal Pradesh, India
[2]M.M. College of Pharmacy, Maharishi Markandeshwar (Deemed to be University), Mullana, Ambala, Haryana, India
[3]Department of Biotechnology, Maharishi Markandeshwar (Deemed to be University), Mullana, Ambala, Haryana, India

Abstract

Dietary flavonoids have emerged as a safe as well as effective way of treating, controlling and managing the various diseases. Various functional foods have proven effective and beneficial for human heath due to presence of flavonoids. Naringin is among the important plant based naturally occurring flavonoids. Citrus fruits are natural source of flavonoid naringin. Naringin is well known for its therapeutic potential. It has been found effective against various diseases. This chapter focuses on naringin, its biological properties and therapeutic potential.

Keywords: flavonoids, naringin, citrus, pharmacokinetics

1. Introduction

Flavonoids comprise the essential domain of secondary metabolites and act as a source of bioactive compounds in plants (Ghasemzadeh & Jaafar 2013). The

---

* Corresponding Author's E-mail: dr.manish.kumar@mmumullana.org.

In: Flavonoids
Editors: Nirmala Sehrawat, Diwakar Aggarwal, Sushil Kumar Upadhyay et al.
ISBN: 978-1-68507-953-6

flavonoids have shown various biological activities that proved beneficial for humans. Literature review provide evidences that probability of myocardial infarction and stroke decreases with increase in dietary intake of flavonoids (Hertog et al., 2012; Keli et al., 1996). These natural compounds have great therapeutic potential for prevention and treatment of various diseases (Benavente-García and Castillo, 2008; Habauzit and Morand, 2012) due to their anti-oxidant, anti-inflammatory, anti-proliferative and anti-thrombotic actions (Bharti et al., 2014; Wright et al., 2013). However, the extent of intracellular modulator actions may vary depending on the gender, lifestyle, disease states and interactions with other drugs thus limiting their impact on human health (Wright et al., 2013; Chanet et al., 2012).

As per Cook and Samman., 1996 approximately 4000 flavonoids were identified and isolated from different fruits, vegetables, and herbs. Croft KD (1998), elucidated that basic flavonoid structure is a polyphenolic compound with 15-carbon atoms, aligned in 3 rings of which 2 are benzene rings connected with a 3-carbon chain in which the A ring develops from resorcinol or phloroglucinol through the acetate pathway and possess distinct hydroxylation pattern at the 5 and 7 position, while the B ring originates from the shikimate pathways. Peterson and Dwyer (1998), classified several types of flavonoids found in plants, such as flavones, flavanones, flavonols, isoflavones, anthocyanidins, and flavanols. Ross and Kasum (2002) proved through in-vivo study that flavonoids are good scavengers of free radicals and prevent oxidative stress.

Naringin, a flavanone glycoside, is a major component of grapefruit juice and responsible for its bitter taste (Peterson et al., 2006). It is also present in citrus fruits (Wong et al., 2013) and imparts a bitter taste to citrus juices (Chtourou et al., 2015). It is one of the main active components of Chinese herbal medicines, such as *Drynaria fortunei* (Kunze) J. Sm. (DF), *Citrus aurantium* L. (CA) and *Citrus medica* L. (CM) (Zhang et al. 2014; Yin et al. 2015).

When ingested, naringin gets metabolized to form several metabolites (including naringenin) in blood and urine. An extensive literature survey has revealed that naringin possesses antioxidant, anti-inflammatory, anti-apoptotic, anti-ulcer, anti-osteoporotic and anti-carcinogenic properties (Wang et al. 2013). However, there have been few reports, until recently, which describe naringin processing.

During study on rats, naringin had shown cardio-protective action in isoproterenol-induced myocardial infarction (Rajadurai and Prince, 2007). In stroke-prone spontaneously hypertensive rats, orally-administered naringin was reported to suppress the age-related increase in blood pressure, to significantly decrease thrombotic tendency and to increase nitric oxide (NO) production thus improving endothelium-dependent vasodilation (Ikemura et al., 2012). Nevertheless, part of these effects could be also probably due to the actions of naringenin, the aglycone formed during the cleavage of the sugar moiety of naringin after its ingestion (Fuhr and Kummert, 1995). Naringenin, by activating mitochondrial BK potassium channels could protect against ischemia-reperfusion injury (Testai et al., 2013). However, there is still the need of investigations about the possible direct cardiovascular physiological actions of naringin. Saponara et al. (2006) showed that naringin could increase the conductance of vascular smooth muscle calcium-activated potassium ion channels (BK(Ca)) but exhibited a poor vasorelaxing action in rat aortic rings compared to its aglycone naringenin. Recently, Saponara et al. (2011) described that a number of flavonoids stimulated or inhibited ICaL in rat tail artery myocytes. They found that naringin and naringenin modestly inhibited ICaL. It has been reported that naringin, at pharmacological relevant concentrations, induced contraction of rat aortic rings and exerted a modest negative inotropic effect on isolated rat hearts (López-Medina et al., 2014). Since naringin could be used as a dietary supplement, antioxidant, anti-inflammatory and even as a template to develop cardiovascular drugs, it is important to investigate its actions on the voltage-dependent Na+ and Ca2+ channels and on the current generated by the Na+ - Ca2+ exchanger in an attempt to elucidate the mechanism of the negative inotropic action of this flavonoid.

2. Dietary Sources of Flavonoid Naringin

Estaquio et al. (2008) and Liu et al. (2004) established that high content of vegetarian diet with fruit and vegetable consumption help in treatment of metabolic syndrome. Martin and Appel (2010) proved that dietary nutrients from plant origin such as polyphenolic compounds (e.g., flavonoids, anthocyanines, and phenolic acids) have potential health benefits for the treatment of obesity, hypertension, cardiovascular diseases, and metabolic syndrome. Among those bioactive compounds, flavonoids constitute a large proportion. Citrus plants such as oranges, mandarins, grapefruit, and acid citrus fruits, namely lemons, bergamots, and limes, have been identified as major source of flavonoids including naringin, naringenin, nobelitin, narirutin, and hesperidin. Tripoli et al. (2007) established that these flavonoids possess strong antioxidant and anti-inflammatory activities both *in vitro* and *in vivo*.

Naringin is a flavanone glycoside which is present in grapes and citrus fruits as major component which is linked with the characteristic bitter taste of grapefruit juice. Two rhamnose units are attached to its aglycon portion, naringenin, at the 7-carbon position. Naringin contents in various citrus species juice are summarized in Table 1. Renugadevi and Prabu (2009) established antioxidant property of naringin and naringenin; however, naringin have low potency in comparison to naringenin because the sugar moiety in the former causes steric hindrance of the scavenging group.

Table 1. Content of Naringin in various citrus species

Fruit name	Naringin content (μg/ml)	Reference
Citrus sinensis	21.3	Ooghe et al., 1994
C. aurantium	19.7	Kawaii et al., 1999
C. reticulata	3383.6	de Lourdes et al., 2007
C clementina	8.0	Dhuique-Mayer et al., 2005
C. bergamia	22.3	Kawaii et al., 1999
C. paradisi	230.0	Kawaii et al., 1999

3. Pharmacokinetics of Flavonoid Naringin

Naringin, a bound form of Naringenin with hexose sugar is absorbed in body after breaking down into aglycone (naringenin) and monosaccharide by gut flora. Kim et al., (1998) identified the micro-flora (Bacteroides JY-6, Eubacterium YK-4, and Fusobacterium K-60) present in the intestine of rat and humans are able to break the glycosidic bond of naringin before being it is absorbed in the body. Choudhary et al. (1999) evaluated the metabolism of naringin which is carried out by the microflora present in the intestine to its aglycon naringenin; which is later absorbed from the gut. Honohan et al. (1976) indicated that intestinal absorption of aglycone flavanones may be greater than 90% then its glycone.

Felgines et al., (2000) checked the bioavailability of naringenin and its glucoside in rat after oral administration of naringenin (0.25%), naringenin-7-glucoside (0.38%) and naringenin-7-rhamnoglucoside and found that Glucurono and sulfoconjugated metabolites of naringenin have no relation with the type or form of naringenin administered. Naringenin-7-rhamnoglucoside showed a delay in absorption in comparison to other 2 forms of naringenin in oral administration studies. During study Davies and Yáñez (2012), observed that the different form of naringenin reached the peak in plasma at different time period and for naringenin it was 20-30 min. (Cmax: 0.021 µmol/L) and for naringenin glucuronide 20-30 min (Cmax: 0.09 µmol/L). In the distribution studies, it was observed that naringenin is widely distributed in the body, starting from the liver at highest concentration followed by stomach, small intestine, kidney, trachea, lung, testis, heart, ovary, spleen, muscle, and brain in succession, however in the brain it was almost negligible indicating its inability to cross the blood-brain barrier.

Some studies performed with the rats gives complete details of the pharmacokinetics of naringin and naringenin. It has been established that naringenin rapidly gets metabolized in the liver to form glucuronide intermediates resulting to limit the bioavailability of naringin in plasma *in vivo* (Fuhr and Kummert, 1995; Ishii et al., 1997; Lee and Reidenberg, 1998). During study, the researchers also reported that a single dose i.v. bolus and oral dose of naringenin and naringin, administered to rats, rapidly underwent conjugation (glucuronoids and sulfates). Wang et al. (2006) observed that the blood serum concentrations of naringenin and naringin sulfates and glucuronides were found almost exclusively in the bloodstream. Moreover, the concentration of naringenin sulfates was much higher than that of naringenin glucuronides. Researchers like Fuhr and Kummert (1995), Ishii et al. (1997), Lee and Reidenberg (1998) also reported the presence of naringenin in urine and plasma (Ishii et al. (1997) and Ameer et. al (1996)) after an oral dose of naringin or grapefruit juice.

As per Manach et al., 2005, Naringenin was also found to be excreted from the urine in the range from 8.5- 8.8% of administrated dose. In human case studies done by Fuhr and Kummert (1995), have shown that the presence of naringenin was detected after 8 hours of administration of naringin. Most administrated part was found to be excreted in urine, which was 57% molar equivalent of administrated naringin, beside naringenin they were also able to detect some other minor components like naringenin glucuronides, (<4 µmol/L).

4. Toxicological Studies

Flavonoids are generally considered to have low toxicity, but in the case of naringenin experimental data is very limited. First time a study done by Wilcox et al., in 1999 showed the IC50 value was > 1mM but it varies cell to cells. Later another study done by Tundis et al. (2011) have shown that IC50 value for the naringenin was 2.2, 7.7, and 33.4 μM in C32, LNCaP, and COR-L23 cell lines respectively. However, as per Zandi et al., (2011), IC50 value for the Vero cells was found to be much higher (0.319 mM). In-vivo toxicity during study by Ortiz et al., (2008) showed that medium lethal dose (LD50) for mice and rat is >5000 mg/kg body weight. While as toxicity to embryonic tissue by Perez-Pasten et al., (2010), of mice was observed at 300 μM conc.

5. Daily Intake of Flavonoid Naringin

Scalbert and Williamson (2000) recommended average daily human intake of polyphenols as ~1 g/day while daily requirement of naringin or flavonoids is not established which but as per Beecher (2003) it may vary depending upon the geographical and cultural changes for example 20 mg/day in United States, Denmark, Finland while 70 mg/day in The Netherlands. Chun et al. (2007) projected flavonoid intake by combining the USDA flavonoids database and 24h dietary recall in NHANES 1999–2002 data. The daily major intake of flavonoids was found from tea (157 mg), citrus fruit juices (8 mg), wine (4 mg), and citrus fruits (3 mg). One more study has been conducted by Lyons-Wall et al. (2004) in the Australian population, estimated a flavonoid intake of 128 mg/d from 15 flavonoids except for isoflavones. The ambiguity of a daily intake value for flavonoids is mainly due to the lack of comprehensive food composition data for >=1 flavonoid subclasses (Beecher, 2003).

6. Biological Properties of Flavonoid Naringin

It has been reported that citrus fruit extract inhibit the cytochrome enzymes or interfere with the uptake process of drugs from intestine thus influences the bioavailability of many drugs (Fuhr and Kummert, 1995; Ameer and Weintraub, 1997; Bailey 2010). Organic anion transporting polypeptides (OATPs) constitute an important family of sodium-independent transport proteins. As per Bailey et al. (2007) and Dresser et al. (2002) grapefruit juice and naringin restrict the organic anion-transporting polypeptide 1A2 (OATP1A2)-mediated fexofenadine absorption and inhibit the oral bioavailability of the drug. Many of the bioactivities of citrus flavonoids appear to affect vascular endothelial cells. Trombetta et al. (2010) reported the potential antioxidant activity of Bergamot (Citrus bergamia Risso), in human umbilical vein endothelial cells, and inhibited the activation of NF-kB.

Chen et al. (2012) reported that water extracts of sweet orange peel prevented the tert-butyl hydroperoxide–induced cytotoxic effect of HepG2 cells at 50- to 500-mg/mL doses and inhibited TBARS generation, increased mitochondrial membrane potential and Bcl-2: Bax ratio, and decreased caspase-3 activation. Sugiura et al. (2006a) observed that citrus fruit extracts offer potential health benefit in diabetes and obesity. Mandarin fruit extracts were also reported to improve the metabolic function of liver and restored the antioxidant enzymes in streptozotocin induced diabetic rats. Sugiura et al. (2006b) reported the potential benefits (improved glucose tolerance) for *Citrus unshiu* extract administration for 10 wk in for the treatment of type 2 diabetic Goto-Kakizaki rats. Insulin resistance in type 2 diabetes may also lead to hepatic steatosis accompanied by progressive inflammation of the liver.

Park et al. (2013) evaluated the antihyperglycemic effect after administration of peel extract of *Citrus unshiu* in male mice. It was observed to inhibit the hepatic gluconeogenic phosphoenol pyruvate carboxykinase mRNA expression and its activity and by inducing insulin/glucagon secretion may have antihyperglycemic effect. These mice also displayed a significant reduction of body weight gain and body fat mass. *Citrus unshiu* extract also improved hepatic steatosis and hypertriglyceridemia by inhibiting gene expression and by increasing the activity of lipogenic enzymes and FA oxidation in the liver. *Citrus unshiu* may have these benefits due to the increased concentrations of anti-inflammatory adiponectin and IL-10, and decreased concentrations of proinflammatory markers [IL-6, monocyte chemotactic protein 1 (MCP-1), IFN-g, and TNF-a] in the plasma or liver.

Similarly, Kang et al. (2012) observed that *Citrus sunki* extract may help in decreasing the body weight gain by reducing the adipose tissue weight, serum total cholesterol, and TG and serum concentrations of aspartate transaminase (AST), alanine transaminase (ALT), and alkaline phosphatase (ALP) in high-fat-diet–fed mice and prevented liver steatosis. Hepato-protective and antilipemic effect of *Citrus sunki* extracts are mediated via upregulation of phosphorylation levels of AMP-activated protein kinase (AMPK) and acetyl-CoA carboxylase (ACC), which are related to FA b-oxidation. Ramful et al. (2010) identified that Citrus fruit extracts inhibit the advanced glycation end product (AGE)– and H_2O_2-induced oxidative stress in human adipocytes.

Fresh grape fruit extracts when evaluated through a randomized clinical study with obese patients, resulted in decreased body weight and improved insulin resistance (Fujioka et al. (2006)) by increasing human adipose lipolysis at a dose of 1.4 g/d, probably by inhibiting cAMP-phosphodiesterase (cAMP-PDE) (Dallas (2008)) or by enhancing phosphorylation of cAMP-dependent protein kinase A (PKA) and hormone-sensitive lipase (HSL) in mature 3T3-L1 adipocytes (Kang et al. (2012)).

7. Therapeutic Potential of Flavonoid Naringin

Naringin, until recently, has been shown to exert potential therapeutic actions by modulating various protein and enzyme expressions as summarized as below:

7.1. Anti-Inflammatory Actions

Naringin is well known for its anti-inflammatory properties (Table 2).

Table 2. Anti-inflammatory properties of naringin

Author	Effect
Habauzit et al. 2011	restrain the production of regulated upon activation normal T-cell expressed and secreted (RANTES) via inhibition of the nuclear translocation of NF-jB in a human epidermal keratinocytes cell line (HaCaT cells)
Tan et al. 2014	In the 20-month-old male Wistar rats, reduced the expression of signallig factors inflammatory like interleukin-6 (IL-6), interleukin-8 (IL-8), inducible nitric oxide synthase (iNOS), nuclear factor erythroid 2-related factor 2 (Nrf2) and TNF-a, in animal models of inflammation.
Liu et al. 2011	inhibited LPS-induced iNOS expression and NO production in macrophages
Luo et al. 2012	In a guinea pig having chronic bronchitis, reduced the concentrations of IL-8 and leukotriene B4 in bronchoalveolar lavage fluid (BALF) and decreased myeloperoxidase activity in both the BALF and lung tissue. It also increased the superoxide dismutase (SOD) activity in lung tissue and enhanced the lipoxin A4 level in BALF.
Gopinath & Sudhandiran 2012	In rats with 3-nitropropionic acid (3-NP)-induced experimental Huntington's disease, reduced 3-NP-induced inflammation through the modulation of Nrf2-driven ARE gene expression and the reduction of TNF-a, COX-2 and iNOS expression. Nrf2-mediated upregulation of gene expression confers the protective effects of naringin in 3-NP-induced inflammation by reducing the deleterious production of pro-inflammatory mediators
Pereira et al. 2007	form a 1:1 complex with Cu (II), in which the Cu (II) ion is coordinated via positions 4 and 5 of naringin. The naringin-Cu (II) complex showed higher anti inflammatory activity than free naringin without reducing cell viability

7.2. Anti-Cancer Effect

Naringin has been shown to have anti-cancer effect (Table 3).

Table 3. Anti-cancer effect of naringin

Author	Effect
Li et al. 2013	Inhibit cell proliferation and to promote cell apoptosis in tumour cells, including triple-negative breast cancer (TNBC) cells, human cervical cancer (SiHa) cells and bladder cancer cells. In TNBC cells, the pro-apoptotic activity of naringin results from G1-phase cell cycle arrest. Suppression of the growth of breast cancer cells by naringin is mediated by inhibition of the b-catenin pathway, leading to a significantly increased p21 level and decreased cell survival.
Ramesh & Alshatwi 2013	Naringin decreases SiHa cell proliferation in a dose-dependent manner in the 0–2000 lM concentration range
Tan et al. 2014	Down regulates vascular cell adhesion molecule-1, which is associated with metastasis, by increasing miR-126 expression
Lewinska et al. 2015	In the DU145 prostate cancer cell line, the 5-methyl-20-deoxycytidine level is decreased after naringin treatment, indicating that it is a DNA hypomethylating agent with potential to modulate gene expression. In the 20–50 lM concentration range, naringin causes a decrease in DNA methyltransferase activity
Kim et al. 2008	inhibits the proliferation of 5637 bladder cancer cells and induces G1 arrest by upregulation of p21

7.3. Effect on Metabolic Syndrome

The effect of naringin on metabolic syndrome has been shown in Table 4.

Table 4. Effect of naringin on metabolic syndrome

Author	Effect
Wang et al. 2015	In mice fed a high fat diet (HFD) for 20 weeks, naringin has been shown to significantly ameliorate mitochondrial dysfunction by restoring the mitochondrial matrix metalloproteinase activity and reactive oxygen species (ROS) and ATP levels in mitochondria
Pu et al. 2012	long-term administration of naringin to mice fed a HFD has been reported to reduce the risk of metabolic syndrome via upregulation of adenosine monophosphate (AMP)-activated protein kinase (AMPK) and its related signalling pathway
Alam et al. 2014	In low-density lipoprotein (LDL) receptor knockout mice, naringin (0.02 g/100 g bw) has been shown to significantly reduce HMG-CoA reductase activity and to increase production of NO metabolites in the urine. In thoracic aortic ring preparations, naringin improved endothelial function via acetylcholine-mediated NO production.
Alam et al. 2013	In rats fed a HFD, naringin improved metabolic syndrome to a similar extent as rutin or quercetin. Naringin enhanced NO production, improved endothelial function and decreased the cerebral thrombotic tendency.
Baskaran et al. 2015	In rats fed high-fat and high cholesterol diet and in rabbits fed high-cholesterol diet, naringin has been demonstrated to ameliorate hypercholesterolemia and atherosclerosis, inhibit HMG-CoA reductase, and decrease plasma cholesterol, LDL, triglycerides, and hepatic lipid levels without altering the high-density lipoprotein cholesterol level.

7.4. Effect on Oxidative Stress

The effect of naringin on oxidative stress has been listed in Table 5.

Table 5. Effect of naringin on metabolic syndrome

Author	Effect
Bacanli et al. 2015	At concentrations of 5–2000 lM, naringin showed antioxidant activity and reduced the frequency of DNA damage by H2O2 in Chinese hamster fibroblast (V79) cells.
Dhanya et al. 2015	In L6 myoblast cells subjected to oxidative stress, pretreatment with naringin (3 or 24 h) reversed the decrease in glutathione (GSH) and increases in both intracellular free radicals and glucose uptake. Naringin (100 lM) also led to a reduction (~40%) in protein glycation
Chen et al. 2015	In the rat glomerular mesangial cell line HBZY-1, naringin (10 lM) obviously activated the Nrf2 signalling pathway and enhanced the expression and activity of its downstream target, HO-1
Chtourou et al. 2015	Treatment with naringin (25, 50 or 100 mg/kg) significantly and dose dependently resisted all of the biochemical and molecular alterations caused by cisplatin in aged rats
Golechha et al. 2014	In pentylenetetrazole-induced seizure rats, naringin reduces oxidative stress by binding free radicals and regulating the GSH level. Pretreatment with naringin for several hours has been shown to significantly attenuate pentylenetetrazole-induced elevations in the brain malondialdehyde and TNF-a level and to conserve GSH
Bakheet & Attia 2011	In diabetic rats, naringin reduced diabetes-induced lipid peroxidation and ROS accumulation in sperm and curbed the reduction in the GSH: oxidized GSH ratio
Mahmoud et al., 2012	In HFD/STZ-induced diabetic rats with hyperglycemia-induced oxidative damage, naringin enhanced the activity of the antioxidant defence system and thereby conferred protection by decreasing the activities of hepatic SOD, GSH reductase, GSH peroxidase and catalase
Kumar et al. 2010	In D-galactose-treated mice, chronic administration of naringin for a period of 6 weeks attenuated oxidative damage via decreases in lipid peroxidation and the nitrite concentration and restoration of the reduced GSH level as well as SOD, catalase and GSH-S-transferase activities

7.5. Effect on Genetic Damage

The effect of naringin on genetic damage has been listed in Table 6.

Table 6. The effect of naringin on genetic damage

Author	Effect
Lewinska et al. 2015	Some reagents and medicines (e.g., H2O2 and lomefloxacin) cause DNA damage that, if unresolved, can lead to genetic mutations and/or genomic instability
Bacanli et al. 2015	In V79 cells, naringin has antioxidant properties and confers protection against H2O2-induced chromosome breakage and loss and DNA damage. Naringin may avoid H2O2-induced oxidative damage not only by decreasing DNA damage but also by increasing the DNA repair capacity

Author	Effect
Attia 2008	Pretreatment with naringin has been shown to attenuate lomefloxacin-induced genomic instability in mice. A dose dependent effect was observed at concentrations of 5–50 mg/kg and naringin significantly reduced cell proliferation, chromosomal aberrations and micronucleus formation in bone marrow and increased mitotic activity
Kumar et al. 2014	In male Wistar rats, naringin has been shown to modulate the production and expression of oxidative mediators and to reduce DNA damage to relieve the symptoms of inflammatory bowel disease
Bakheet & Attia 2011	In diabetic rats, administration of naringin for 4 weeks significantly decreased the rate of DNA strand breaks. The anti-genotoxic effect of naringin is mediated by the inhibition of hyperglycemia-induced generation of free radicals

7.6. Effect on CNS Diseases

The effect of naringin on CNS has been listed in Table 7.

Table 7. The effect of naringin on CNS

Author	**Effect**
Sachdeva et al. 2014	In rats treated intracerebroventricularly with streptozotocin (ICV-STZ rats), it has been shown to result in significant increases of the malondialdehyde and nitrite levels and marked reductions in the GSH level and enzymatic activities of SOD and catalase in the brain. Naringin also results in the suppression of acetylcholinesterase activity and TNF-a levels in the brains of ICV-STZ rats
Jung & Kim 2014	In a rat model of Parkinson's disease, intraperitoneal injection of naringin protects the nigrostriatal dopaminergic (DA) projection by increasing glial cell line-derived neurotrophic factor expression and decreasing TNF-a expression in DA neurons and microglia
Jeong et al. 2015	In kainic acid (KA)-treated mice, naringin treatment significantly reduced seizure. Moreover, it protected hippocampal CA1 neurons in the KA-treated hippocampus, ameliorated KA-induced autophagic stress, decreased microtubule-associated protein light chain 3 (LC3) expression and attenuated an improvement in the TNF-a level

Conclusion and Future Prospective

Dietary flavonoids have emerged as a safe and effective way of treating; controlling or managing the various diseases. Various functional foods have proven effective and beneficial for human heath due to presence of flavonoids. Naringin is a plant based naturally occurring flavonoid. Citrus fruits are natural source of flavonoid naringin. Naringin is well known for its therapeutic potential. It has been found effective against various diseases. Naringin has been reported for its antimicrobial, anti-inflammatory, anti-oxidant, anti-diabetic and anti-cancerous properties. It has been found to have hepatoprotective, reno-protective and cardio-protective property. Also, effect of naringin on other various diseases including osteogenesis, controlled medication discharge, bone recovery, osteoporosis, osteoblasts, metabolic condition, chromosomal damage, oxidative damage and sensory system issue have been reported so far and found promising. Further studies and clinical evidences are required to establish the naringin based therapy for any particular disease. Naringin may provide effective, safe and alternate way to manage various diseases.

References

Alam MA, Kauter K, Brown L. 2013. Naringin improves diet-induced cardiovascular dysfunction and obesity in high carbohydrate, high fat diet-fed rats. *Nutrients.* 5:637–650.

Alam MA, Subhan N, Rahman MM, Uddin SJ, Reza HM, Sarker SD. 2014. Effect of citrus flavonoids, naringin and naringenin, on metabolic syndrome and their mechanisms of action. *Adv Nutr.* 5:404–417.

Ameer B, Weintraub RA, Johnson JV, Yost RA, Rouseff RL. (1996) Flavanone absorption after naringin, hesperidin, and citrus administration. *Clin Pharmacol Ther;* 60:34–40.

Amic D, Davidovic-Amic D, et al. (2003) Structure radical scavenging activity relationships of flavonoids, *Croat Chem Act,* 76(18): 55-61.

Attia SM. 2008. Abatement by naringin of lomefloxacin-induced genomic instability in mice. *Mutagenesis.* 23:515–521.

Bailey DG, Dresser GK, Leake BF, Kim RB. (2007) Naringin is a major and selective clinical inhibitor of organic anion-transporting polypeptide 1A2 (OATP1A2) in grapefruit juice. *Clin Pharmacol Ther;* 81: 495–502.

Bacanli M, Basaran AA, Basaran N. 2015. The antioxidant and antigenotoxic properties of citrus phenolics limonene and naringin. *Food Chem Toxicol.* 81:160–170.

Bakheet SA, Attia SM. 2011. Evaluation of chromosomal instability in diabetic rats treated with naringin. *Oxid Med Cell Longev.* 2011:365292.

Baskaran G, Salvamani S, Ahmad SA, Shaharuddin NA, Pattiram PD, Shukor MY. 2015. HMG-CoA reductase inhibitory activity and phytocomponent investigation of Basella alba leaf extract as a treatment for hypercholesterolemia. *Drug Des Devel* Ther. 9:509–517.

Beecher GR. (2003) Overview of dietary flavonoids: nomenclature, occurrence and intake. *J Nutr* 133 Suppl:3248S–54S.

Bharti S, Rani N, Krishnamurthy B, Arya DS (2014) Preclinical evidence for the pharmacological actions of naringin: A review. *Planta Med* 80: 437-451.

Chanet A, Milenkovic D, Manach C, Mazur A, Morand C (2012) Citrus flavanones: what is their role in cardiovascular protection? *J Agric Food Chem* 60: 8809-8822.

Chen Z-T, Chu H-L, Chyau C-C, Chu C-C, Duh P-D. (2012) Protective effects of sweet orange (Citrus sinensis) peel and their bioactive compounds on oxidative stress. *Food Chem;* 135:2119–27.

Chen F, Zhang N, Ma X, Huang T, Shao Y, Wu C, Wang Q. 2015. Naringin alleviates diabetic kidney disease through inhibiting oxidative stress and inflammatory reaction. *PLoS One.* 10:e0143868.

Choudhury R, Chowrimootoo G, Srai K, Debnam E, Rice-Evans CA. (1999) Interactions of the flavonoid naringenin in the gastrointestinal tract and the influence of glycosylation. *Biochem Biophys Res Commun* 265:410–5.

Chtourou Y, Gargouri B, Kebieche M, Fetoui H. (2015) Naringin abrogates cisplatin-induced cognitive deficits and cholinergic dysfunction through the down-regulation of AChE expression and iNOS signaling pathways in hippocampus of aged rats. *J Mol Neurosci.* 56:349–362

Chun OK, Chung SJ, Song WO. (2007) Estimated dietary flavonoid intake and major food sources of U.S. adults. *J Nutr;*137:1244–52.

Cook NC, Samman S. (1996) Flavonoids—chemistry, metabolism, cardioprotective effects, and dietary sources. *J Nutr Biochem;*7: 66–76.

Croft KD. (1998) The chemistry and biological effects of flavonoids and phenolic acids. *Ann N Y Acad Sci;*854:435–42.

Dallas C, Gerbi A, Tenca G, Juchaux F, Bernard F-X. (2008) Lipolytic effect of a polyphenolic citrus dry extract of red orange, grapefruit, orange (SINETROL) in human body fat adipocytes: mechanism of action by inhibition of cAMP-phosphodiesterase (PDE). *Phytomedicine;* 15: 783–92.

Davies NM, Yáñez JA. (2012) *Flavonoid Pharmacokinetics: Methods of Analysis, Preclinical and Clinical Pharmacokinetics, Safety, and Toxicology.* wiley publication, New Jersey, 1-352.

de Lourdes Mata Bilbao M, Andrés-Lacueva C, Jáuregui O, Lamuela- Raventós RM. (2007) Determination of flavonoids in a citrus fruit extract by LC–DAD and LC–MS. *Food Chem;* 101:1742–7.

Dhanya R, Arun KB, Nisha VM, Syama HP, Nisha P, Santhosh Kumar TR, Jayamurthy P. 2015. Preconditioning L6 muscle cells with naringin ameliorates oxidative stress and increases glucose uptake. *PLoS One.* 10: e0132429.

Dhuique-Mayer C, Caris-Veyrat C, Ollitrault P, Curk F, Amiot MJ. (2005) Varietal and interspecific influence on micronutrient contents in citrus from the Mediterranean area. *J Agric Food Chem;* 53:2140–5.

Dresser GK, Bailey D, Leake B, Schwarz U, Dawson P, Freeman D, Kim R. (2002) Fruit juices inhibit organic anion transporting polypeptide–mediated drug uptake to decrease the oral availability of fexofenadine. *Clin Pharmacol Ther;* 71:11–20.

Estaquio C, Castetbon K, Kesse-Guyot E, Bertrais S, Deschamps V, Dauchet L, Péneau S, Galan P, Hercberg S. (2008) The French National Nutrition and Health Program Score is associated with nutritional status and risk of major chronic diseases. *J Nutr* 138:946–53.

Felgines C, Texier O et al. (2000) Bioavailability of the flavanone naringenin and its glycosides in rats. *American Journal of Physiology—Gastrointestinal and Liver Physiology,* 279(6): G1148–G1154.

Fuhr U, Kummert AL (1995) The fate of naringin in humans: A key to grapefruit juice-drug interactions? *Clin Pharmacol Ther* 58: 365-373.

Fujioka K, Greenway F, Sheard J, Ying Y. (2006) The effects of grapefruit on weight and insulin resistance: relationship to the metabolic syndrome. *J Med Food;* 9:49–54.

Ghasemzadeh A, Jaafar HZ. (2013) Profiling of phenolic compounds and their antioxidant and anticancer activities in pandan (Pandanus amaryllifolius Roxb.) extracts from different locations of Malaysia. *BMC Complement Altern Med.* 13:341.

Golechha M, Sarangal V, Bhatia J, Chaudhry U, Saluja D, Arya DS. 2014. Naringin ameliorates pentylenetetrazol-induced seizures and associated oxidative stress, inflammation, and cognitive impairment in rats: possible mechanisms of neuroprotection. *Epilepsy Behav.* 41:98–102

Gopinath K, Sudhandiran G. 2012. Naringin modulates oxidative stress and inflammation in 3-nitropropionic acid-induced neurodegeneration through the activation of nuclear factor-erythroid 2-related factor-2 signalling pathway. *Neuroscience.* 227:134–143.

Habauzit V, Sacco SM, Gil-Izquierdo A, Trzeciakiewicz A, Morand C, Barron D, Pinaud S, Offord E, Horcajada MN. 2011. Differential effects of two citrus flavanones on bone quality in senescent male rats in relation to their bioavailability and metabolism. *Bone.* 49:1108–1116.

Habauzit V, Morand C (2012) Evidence for a protective effect of polyphenols-containing foods on cardiovascular health: an update for clinicians. *Ther Adv Chronic Dis* 3: 87-106.

Honohan T, Hale RL, et al. (1976) Synthesis and metabolic fate of hesperetin-3-14C, *J Agricul Food Chem,* 24(5), 906–911.

Ikemura M, Sasaki Y, Giddings JC, Yamamoto J (2012) Preventive effects of hesperidin, glucosyl hesperidin and naringin on hypertension and cerebral thrombosis in stroke-prone spontaneously hypertensive rats. *Phytother Res* 26: 1272-1277.

Jung UJ, Kim SR. 2014. Effects of naringin, a flavanone glycoside in grapefruits and citrus fruits, on the nigrostriatal dopaminergic projection in the adult brain. *Neural Regen Res.* 9:1514–1517.

Jeong KH, Jung UJ, Kim SR. 2015. Naringin attenuates autophagic stress and neuroinflammation in kainic acid-treated hippocampus in vivo. *Evid Based Complement Alternat Med.* 2015:354326. doi:10.1155/2015/354326.

Kang SI, Shin HS, Kim HM, Hong YS, Yoon SA, Kang SW, Kim JH, Kim MH, Ko HC, Kim SJ. (2012) Immature Citrus sunki peel extract exhibits antiobesity effects by beta-oxidation and lipolysis in high-fat diet-induced obese mice. *Biol Pharm Bull;* 35:223–30.

Kawaii S, Tomono Y, Katase E, Ogawa K, Yano M. (1999) HL-60 differentiating activity and flavonoid content of the readily extractable fraction prepared from citrus juices. *J Agric Food Chem;* 47:128–35.

Keli SO, Hertog MGL, Feskens EJM, Kromhout D (1996) Dietary flavonoids, antioxidant vitamins, and the incidence of stroke. *Arch Int Med* 156: 637-642.

Kim DH, Jung EA, et al. (1998) Intestinal bacterial metabolism of flavonoids and its some biological activities, *Arch Pharm Res,* 21(1): 17–23.

Kim DI, Lee SJ, Lee SB, Park K, Kim WJ, Moon SK. 2008. Requirement for Ras/Raf/ERK pathway in naringin-induced G1-cell-cycle arrest via p21WAF1 expression. *Carcinogenesis.* 29:1701–1709.

Kumar A, Prakash A, Dogra S. 2010. Naringin alleviates cognitive impairment, mitochondrial dysfunction and oxidative stress induced by D-galactose in mice. *Food Chem Toxicol.* 48:626–632.

Kumar VS, Rajmane AR, Adil M, Kandhare AD, Ghosh P, Bodhankar SL. 2014. Naringin ameliorates acetic acid induced colitis through modulation of endogenous oxido-nitrosative balance and DNA damage in rats. *J Biomed Res.* 28:132–145.

Lewinska A, Siwak J, Rzeszutek I, Wnuk M. 2015. Diosmin induces genotoxicity and apoptosis in DU145 prostate cancer cell line. *Toxicol in Vitro.*29:417–425.

Li H, Yang B, Huang J, Xiang T, Yin X, Wan J, Luo F, Zhang L, Li H, Ren G. 2013. Naringin inhibits growth potential of human triple-negative breast cancer cells by targeting beta-catenin signaling pathway. *Toxicol Lett.* 220:219–228.

Liu Y, Wu H, Nie YC, Chen JL, Su WW, Li PB. 2011. Naringin attenuates acute lung injury in LPS-treated mice by inhibiting NF-jB pathway. *Int Immunopharmacol.* 11:1606–1612.

López-Medina AI, Alvarez-Collazo J, Rodríguez AA, Morón-Rodríguez F, Cabrera-Suárez H, Alvarez JL (2014) Direct actions of naringin on rat cardiac and vascular smooth muscle. *Bol Latinoam Caribe Plant Med Aromat* 13: 238-248.

Luo YL, Zhang CC, Li PB, Nie YC, Wu H, Shen JG, Su WW. 2012. Naringin attenuates enhanced cough, airway hyperresponsiveness and airway inflammation in a guinea pig model of chronic bronchitis induced by cigarette smoke. *Int Immunopharmacol.* 13:301–307.

Lyons-Wall P, Autenzio P, Lee E, Moss R, Samman S. (2004) Catechins are the major source of flavonoids in a group of Australian women. *Asia Pac J Clin Nutr.* 13:S72.

Mahmoud AM, Ashour MB, Abdel-Moneim A, Ahmed OM. 2012. Hesperidin and naringin attenuate hyperglycemia-mediated oxidative stress and proinflammatory cytokine production in high fat fed/streptozotocin-induced type 2 diabetic rats. *J Diabetes Complications.* 26:483–490.

Manach C, Williamson G, et al. (2005) Bioavailability and bioefficacy of polyphenols in humans. I. Review of 97 bioavailability studies, *Am J Clin Nutr,* 81(suppl. 1): 230S–242S.

Martin K, Appel C. (2010) Polyphenols as dietary supplements: A doubleedged sword. *Nutrition and Dietary Supplements,* 2:1–12.

Ooghe WC, Ooghe SJ, Detavernier ClM, Huyghebaert A., (1994) Characterization of orange juice (Citrus sinensis) by flavanone glycosides. *J Agric Food Chem;* 42:2183–90.

Ortiz-Andrade RR, Sanchez-Salgado, et al. (2008) Antidiabetic and toxicological evaluations of naringenin in normoglycaemic and NIDDM rat models and its implications on extra-pancreatic glucose regulation, *Diabetes Obes Metab,* 10(11): 1097–1104.

Park H-J, Jung UJ, Cho S-J, Jung H-K, Shim S, Choi M-S. (2013) Citrus unshiu peel extract ameliorates hyperglycemia and hepatic steatosis by altering inflammation and hepatic glucose- and lipid-regulating enzymes in db/db mice. *J Nutr Biochem;* 24:419–27.

Pereira RM, Andrades NE, Paulino N, Sawaya AC, Eberlin MN, Marcucci MC, Favero GM, Novak EM, Bydlowski SP. 2007. Synthesis and characterization of a metal complex containing naringin and Cu, and its antioxidant, antimicrobial, antiinflammatory and tumor cell cytotoxicity. *Molecules.* 12:1352–1366.

Perez-Pasten R, Martinez-Galero E, et al. (2010) Quercetin and naringenin reduce abnormal development of mouse embryos produced by hydroxyurea, *J Pharm Pharmacol,* 62(8): 1003-1009.

Peterson JJ, Beecher GR, Bhagwat SA, Dwyer JT, Gebhardt SE, Haytowitz DB, Holden JM (2006) Flavanones in grapefruit, lemons, and limes: A compilation and review of the data from the analytical literature. *J Food Comp Anal* 19: S74-S80.

Peterson J, Dwyer J. (1998) Flavonoids: dietary occurrence and biochemical activity. *Nutr Res;*18:1995–2018.

Pu P, Gao DM, Mohamed S, Chen J, Zhang J, Zhou XY, Zhou NJ, Xie J, Jiang H. 2012. Naringin ameliorates metabolic syndrome by activating AMP-activated protein kinase in mice fed a high-fat diet. *Arch Biochem Biophys.* 518:61–70.

Rajadurai M, Prince PS (2007) Preventive effect of naringin on isoproterenol-induced cardiotoxicity in Wistar rats: an in vivo and in vitro study. *Toxicology* 232: 216-225.

Ramesh E, Alshatwi AA. 2013. Naringin induces death receptor and mitochondria-mediated apoptosis in human cervical cancer (SiHa) cells. *Food Chem Toxicol.* 51:97–105.

Ramful D, Tarnus E, Rondeau P, Robert Da Silva C, Bahorun T, Bourdon E. (2010) Citrus fruit extracts reduce advanced glycation end products (AGEs) and H2O2-induced oxidative stress in human adipocytes. *J Agric Food Chem.*

Renugadevi J, Prabu SM. (2009) Naringenin protects against cadmium induced oxidative renal dysfunction in rats. *Toxicology* 256: 128–34.

Ross JA, Kasum CM. (2002) Dietary flavonoids: bioavailability, metabolic effects, and safety. *Annu Rev Nutr;* 22:19–34.

Sachdeva AK, Kuhad A, Chopra K. 2014. Naringin ameliorates memory deficits in experimental paradigm of Alzheimer's disease by attenuating mitochondrial dysfunction. *Pharmacol Biochem Behav.* 127:101–110.

Saponara S, Testai L, Iozzi D, Martinotti E, Martelli A, Chericoni S, Sgaragli G, Fusi F, Calderone V (2006) (+/-)-Naringenin as large conductance Ca2+-activated K+ (BKCa) channel opener in vascular smooth muscle cells. *Br J Pharmacol* 149: 1013-1021.

Saponara S, Carosati E, Mugnai1 P, Sgaragli G, Fusi F (2011) The flavonoid scaffold as a template for the design of modulators of the vascular Cav1.2 channels. *Br J Pharmacol* 164: 1684-1697.

Scalbert A, Williamson G. (2000) Dietary intake and bioavailability of polyphenols. *J Nutr;*130 Suppl: 2073S–85S.

Sugiura M, Ohshima M, Ogawa K, Yano M. (2006a) Chronic administration of Satsuma mandarin fruit (Citrus unshiu Marc.) improves oxidative stress in streptozotocin-induced diabetic rat liver. *Biol Pharm Bull;* 29:588–91.

Sugiura M, Ogawa K, Yano M. (2006b) Effect of chronic administration of fruit extract (Citrus unshiu Marc.) on glucose tolerance in GK rats, a model of type 2 diabetes. *Biosci Biotechnol Biochem;* 70:293–5.

Tan TW, Chou YE, Yang WH, Hsu CJ, Fong YC, Tang CH. 2014. Naringin suppress chondrosarcoma migration through inhibition vascular adhesion molecule-1 expression by modulating miR-126. *Int Immunopharmacol.* 22:107–114.

Testai L, Martelli A, Marino A, D'Antongiovanni V, Ciregia F, Giusti L, Lucacchini A, Chericoni S, Breschi MC, Calderone V (2013) The activation of mitochondrial BK potassium channels contributes to the protective effects of naringenin against myocardial ischemia/reperfusion injury. *Biochem Pharmacol* 85: 1634-1643.

Tripoli E, Guardia ML, Giammanco S, Majo DD, Giammanco M. (2007) Citrus flavonoids: molecular structure, biological activity and nutritional properties: a review. *Food Chem* 104:466–79.

Trombetta D, Cimino F, Cristani M, Mandalari G, Saija A, Ginestra G, Speciale A, Chirafisi J, Bisignano G, Waldron K, et al. (2010) In vitro protective effects of two extracts from bergamot peels on human endothelial cells exposed to tumor necrosis factor-a (TNF-a). *J Agric Food Chem;* 58:8430–6.

Tundis R, Loizzo MR, et al. (2011) In vitro cytotoxic activity of extracts and isolated constituents of Salvia leriifolia Benth. against a panel of human cancer cell lines, *Chem Biodivers,* 8(6), 1152-62.

Wang M, Chao P, Hou Y, Hsiu S,Wen K, Tsai S. (2006) Pharmacokinetics and conjugation metabolism of naringin and naringenin in rats after single dose and multiple dose administrations. *J Food Drug Anal.;* 14: 247–53.

Wang DM, Yang YJ, Zhang L, Zhang X, Guan FF, Zhang LF. (2013) Naringin enhances CaMKII activity and improves long-term memory in a mouse model of Alzheimer's disease. *Int J Mol Sci.* 14:5576–5586.

Wang D, Yan J, Chen J, Wu W, Zhu X, Wang Y. 2015. Naringin improves neuronal insulin signaling, brain mitochondrial function, and cognitive function in High-Fat Diet-induced obese mice. *Cell Mol Neurobiol.* 35:1061–1071.

Wilcox LJ, Borradaile NM, et al. (1999) Antiatherogenic Properties of Naringenin, a Citrus Flavonoid, *Cardiovascular Drug Reviews,* 17(2): 160–178.

Wong KC, Pang WY, Wang XL, Mok SK, Lai WP, Chow HK, Leung PC, Yao XS, Wong MS. (2013) Drynaria fortunei-derived total flavonoid fraction and isolated compounds exert oestrogen-like protective effects in bone. *Br J Nutr.* 110:475–485.

Zandi K, Teoh BT, et al. (2011) In vitro antiviral activity of Fisetin, Rutin and Naringenin against Dengue virus type-2, *Journal of Medicinal Plants Research,* 8(6):, 5534–5539.

Zhang J, Gao W, Liu Z, Zhang Z, Liu C. (2014) Systematic analysis of main constituents in rat biological samples after oral administration of the methanol extract of Fructus aurantii by HPLC-ESI-MS/MS. *Iran J Pharm Res.* 13:493–503.

Chapter 4

Silymarin: A Flavonoid with Diverse Medical Properties

Diwakar Aggarwal[1,*], Gayatri Vashishth[1], Vasudha Datta[1], Chadetrik Rout[2] and Manish Kumar[3]

[1]Department of Biotechnology, Maharishi Markandeshwar (Deemed to be University), Mullana-(Ambala), Haryana, India

[2]Department of Civil Engineering, Maharishi Markandeshwar (Deemed to be University), Mullana- Ambala, Haryana, India

[3]M.M. College of Pharmacy, Maharishi Markandeshwar (Deemed to be University), Mullana, Ambala, Haryana, India

Abstract

Milk thistle (*Silybum marianum*) is a medicinal plant that has been used for thousands of years as a remedy for a variety of ailments. The main component of *S. marianum* fruit extract is silymarin, a flavonolignan which is also known as silybin. Among the several potential biological activities, hepatoprotective activity is the main thing for which silymarin compound is used worldwide. Besides heptoprotectant, it is also beneficial for other disorders such as renal protection, hypolipidemic, cardiovascular protection, anti-oxidant, in neurological disorders, prevention of insulin resistance, cancer and Alzheimer prevention. It is also used as a food remedy and dietary supplement. Present chapter aims to explain the various therapeutic potentials of silymarin along with its chemistry and mechanism of action.

* Corresponding Author's Email: diwakaraggarwal@yahoo.co.in.

In: Flavonoids
Editors: Nirmala Sehrawat, Diwakar Aggarwal, Sushil Kumar Upadhyay et al.
ISBN: 978-1-68507-953-6

Keywords: flavonolignans, galactogogue, heptoprotection, milk thistle, *Silybum marianum*

1. Introduction

Silybum marianum (L.) Gaernt., (synonym *Carduus marianus* L.), belonging to family Asteraceae is hardy and often invasive plant. It is reported to be indigenous to the Mediterranean region whose seeds (also known as achenes), have been valued for their vast medicinal properties (Ladas et al., 2005). It is most commonly known as milk thistle, Marian thistle, Mary's thistle, Saint Mary's thistle, Blessed milk thistle, Mediterranean milk thistle, variegated thistle. The genus name *Silybum*, is attributed to Dioscorides, the first-century Greek physician who gave the name to a number of edible thistles. This plant can be found in North America, Canada, Mexico and Kashmir in India with large leaves and a reddish-purple flower that are all thorny and the medicinal part of the plant is either the seeds or fruits (Figure 1) (Karimi et al., 2011). *S. marianum* fruits vary from milky white to nearly black as they develop toward maturity. There are about 42 fruits/g. A fruit head can produce from 100 to 190 fruits. Silymarin content in the fruits varies between 1.5–3.0%. Fruits containing 3–6% are considered of high quality. Extract prepared from the fruits is standardized, so the silymarin concentration is 70–80% of the extract weight (AbouZid and Ahmad, 2013).

A B

Figure 1. A. *Silybum marianum* (Milk Thistle plant) B. Flower head of *Silybum marianum*.

The specific species name *marianum* L. preserves the legend that the white mottling surrounding the venation of the leaves was a result of a single drop of the Virgin Mary's milk, as it gently fell onto the leaf surface. This medicinal plant is traditionally used as a galactogogue (stimulating milk production), perhaps contributing to the word origins of the common name (Woodward, 1994).

Historically milk thistle is used as heptoprotectant. Pliny the Elder (AD 23–79) reported that the plant had the ability to remove bile. Even Nicolas Culpepper (an herbalist; 1616–54) and Albrecht von Haller (a Swiss physician; 1708–77) mentioned the efficacy of the plant in preventing liver disorders (Mayer et al., 2005). One of the most widely used application of the phytomedicinal extract of milk thistle seeds had been directed toward disorders of the liver, particularly cirrhosis, hepatitis, and in protection and treatment of xenobiotic-mediated liver injury. Beside treatment of liver disorders, it is also used for gall bladder–related disorders, cancer, inflammation and neuro degeneration disorders, hypotensive, styptic, diuretic, and general tonifier (Kroll et al., 2007).

Extracts from *S. marianum* is used in different forms like capsules, powders and homemade tea, however it is important to mention that adverse effects have been reported regarding the consumption of *S. marianum*. A case of anaphylactic shock was reported in a man (54 years) 20 min after consuming a tea containing *S. marianum* tea in 1990 (Geier et al., 1990). After this case, no additional information about *S. marianum* side effects has been reported.

Recently milk thistle is used in oncology, it is being explored for its ability to exert direct cytotoxic effects, mitigate toxicity of certain anticancer agents, and enhance the efficacy of chemotherapeutic agents. Researchers are actively investigating its role as a chemo preventive agent and possible treatment of cancer (Tuli et al., 2021). Several reports showed that milk thistle is a potential cancer chemo-preventive agent (Bai et al., 2018; Si et al., 2019; Tuli et al., 2021).Based on its Medical properties, silymarin-was classified by the WHO in the 1970s as an official medicine with hepatoprotective properties (Wesołowska et al., 2007).

2. Chemistry and Mechanism of Action of Silymarin

Flavonoids are a broad class of secondary metabolites contained in most plants and are responsible for the yellow, red or blue pigments in their flowers. These important plant pigments are also the most common secondary metabolites in the human diet. Because of their beneficial biological effects and their dietary uptake has increased in recent years (Sheikh et al., 2021). Among flavonoids, flavono-lignans are a relatively small subclass of compounds, where the flavonoid part of the molecule is fused with a lignan. Flavonolignans were first discovered in the seeds of *S. marianum* (Figure 1) and silybin is by far the most researched flavonolignan (Biedermann et al., 2014).

Large amounts of Milk thistle seeds are processed to silymarin by the pharmaceutical industry. The term silymarin was introduced in 1968 by the German phytochemist Hildebert Wagner (Wagner et al., 1968) who described the mixture of flavonolignans that had been characterized at the time (silybin, isosilybin silychristin). This mixture was resolved into seven flavonolignans and one flavonoid that comprise 65–80% of milk thistle extract. The milled or pressed seeds (botanically fruits - cypsela - containing only one seed) are extracted with ethanol, methanol, acetone or ethyl acetate. Lipids and polar impurities are removed from the extract and the resulting dry mixture of flavonolignans is called silymarin (Šimanek et al., 2000) (Figure 2). Silymarin has as a molecular formula of $C_{25}H_{22}O_{10}$ and a molecular weight of 482.441, CAS No. 22888-70-6. The silybin structure consists in two main units. The first is based on a taxifolin, which is a flavononol group in flavonoids. The second is a phenyl propanoid unit, which in this case is conyferil alcohol. These two units are linked together into one structure by an oxeran ring (Althagafy et al., 2013). Except silybin A and silybin B, silymarin contains another five major compounds – taxifolin, isosilybin A, isosilybin B, silychristin and silydianin (Figure 3). Silymarin also contains about 30% of undefined polyphenolic compounds often referred to as a polymeric fraction. It is poorly soluble in polar solvents (EtOH and MeOH), and insoluble in non-polar solvents (chloroform and petroleum ether), but highly soluble in polar aprotic solvents such as DMSO, acetone, DMF, and THF (Biedermann et al., 2014).

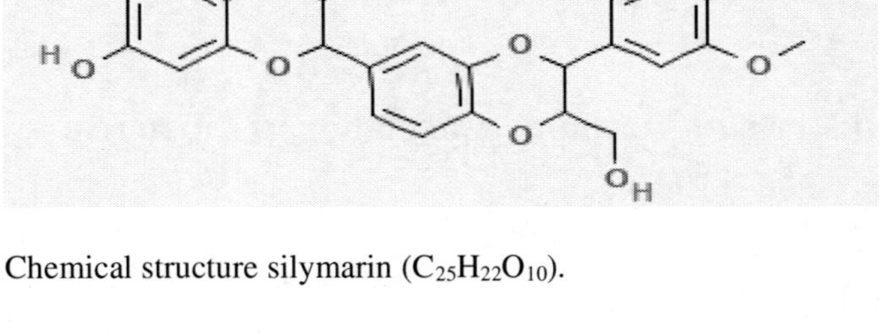

Figure 2. Chemical structure silymarin ($C_{25}H_{22}O_{10}$).

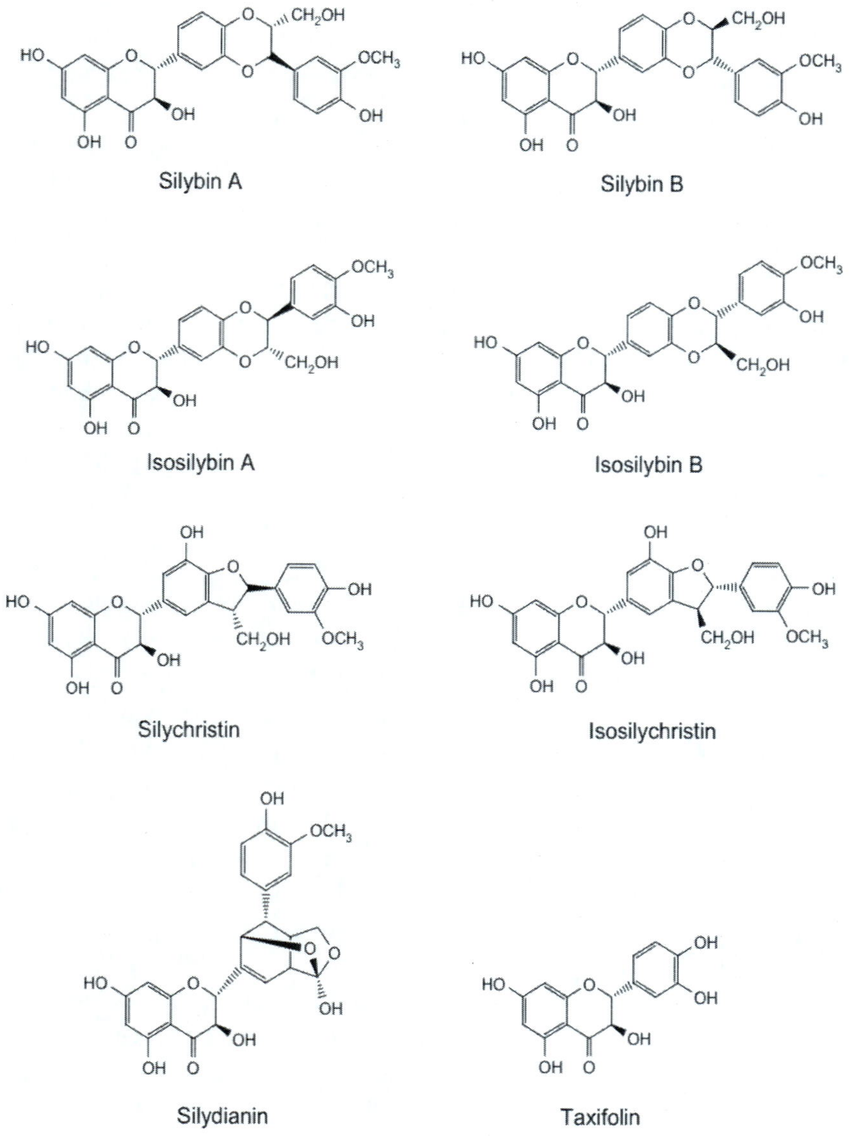

Figure 3. Structure of main silymarin compounds from milk thistle.

In modern therapy, standardized silymarin products are used. According to the European Pharmacopoeia, the nominal silymarin content of the refined and quantified milk thistle dry extract is within the range of 30–65%. The content of silymarin corresponds to sum of the contents of silychristin and silydianin 20-45%, sum of the contents of silybin A and silybin B 40-65%, sum of the contents of isosilybin A and isosilybin B 10-20%. The extract should contain 90-110% of the normal content of silymarin, expressed as silybin (European Pharmacopoeia, 2010). The flavonolignan, silybin, which represents 60-70% of silymarin, has been identified as the major biologically active component (Sonnenbichler, 1986). Silymarin should therefore not be considered as an "alternative medicine" nowadays, as its composition is now chemically well defined.

It has been reported that silymarin prevents lipid peroxidation by scavenging free radicals and increases the levels of reduced glutathione (GSH). It regulates membrane permeability and membrane stability in damage caused by xenobiotics, regulates nuclear expression and inhibits the transformation of hepatocytes in myofibroblasts (with cirrhosis) (Crocenzi and Roma, 2006). New *in vivo* studies have proved silymarin to be a strong hepatoprotective in hepatic fibrosis induced by carbon tetrachloride (CCl_4). After silymarin had been administered to the rats treated previously with CCl_4, the levels of glutamic oxaloacetic transaminase and glutamic pyruvic transaminases were normalized and the secretion of the connective tissue growth factor was decreased, indicating that silymarin ameliorates the destructive changes in the liver after hepatic fibrosis was provoked (Kvasnička et al., 2003). The anticancer properties of silibinin are determined by a variety of cellular pathways which include induction of apoptosis, cell cycle arrest, inhibition of angiogenesis and metastasis, induction of autophagy, induction of anti-oxidant response, alteration in tumor microenvironment by modulating inflammatory response, and synergism with existing anti-cancer drugs (Tuli et al., 2021). An insight has recently been published into the possible mechanism of silymarin defense capacity, as an antioxidant against various hepato-toxicants (Bijak, 2017).

An important challenge in the manufacturing of silymarin-based products is the poor aqueous solubility of flavonolignans which leads to relatively poor intestinal absorption, efforts have been made to increase its oral bioavailability by combining its active component, silibinin, with phosphatidylcholine (compound dB 1016) with liposomes containing variables amount of cholesterol and phospholipids (Maheshwari et al., 2003), or with lipid microspheres formed by an internal oily core, surfactants (e.g., soybean lecithin) and different co-surfactants (Span® 20, Tween® 20 or 80, and propylene glycol) (Abrol et al., 2004). Another formulation that was created to increase silybin bioavailability after oral administration is a compilation of silybin, phosphatidyl choline, and vitamin E, which considerably enhanced silybin's solubility (Di Sario et al., 2005). All these formulations show higher oral bioavailability and facilitate passive targeting to the liver, thus conferring greater pharmacological activity compared with pure silymarin.

3. Therapeutic Potentials of Milk Thistle

Silymarin has been known from many years as a natural remedy for a variety of diseases, various beneficial effects of *S. marianum* or its derivatives have been reported including hepatoprotection, renal protection, hypolipidemic and anti-atherosclerosis activities, cardiovascular protection, prevention of insulin resistance, especially in cirrhotic patients, cancer, and Alzheimer prevention. Silymarin in the herbal medicine is consumed worldwide to treat different types of diseases and liver disorders (Murata et al., 2010).

3.1. Hepatoprotective Activity

Silymarin has been used for many centuries to treat a variety of liver and gallbladder disorders, including hepatitis, cirrhosis and jaundice. It has been used for the treatment of numerous liver disorders characterized by functional impairment or degenerative necrosis. Its hepatoprotective activity is unique and acts in different ways, including antioxidant and anti-inflammatory activities, cell permeability regulator and membrane stabilizer, stimulation of liver regeneration and inhibition of deposition in collagen fibers, which may lead to cirrhosis (Bahmani et al., 2015). Different clinical and laboratory studies have also documented that silymarin protects the liver against toxicities resulting from various chemical toxins such as carbon tetrachloride, acetaminophen, and tetrachloromethane (Šimanek et al., 2000) as well as and environmental toxins, such as snake and insect bites, mushroom and alcohol poisoning (Mastron et al., 2015).

Numerous studies have strongly suggested that the hepatoprotective effects of silymarin are mainly due to its antioxidant activity and free radical scavenging property. This effect is reflected by glutathione modulation and the membrane stabilization that it produces (Yang et al., 2014). In one of the studies conducted by Song et al., (2006) demonstrated that silymarin (200 mg/kg), which was induced by administering ethanol through gavage at a dosage of 5 g/kg body weight every 12 h for a total of three doses, reduced oxidative stress in model mice with alcoholic liver disease. Silymarin treatment also prevented increases in alanine aminotransferase and TNF- α levels as well as decreases in glutathione and lipid peroxidation. It was further reported that silymarin works by reducing metabolic activation by carbon tetrachloride and by acting as an antioxidant that prevents chain rupture (Lettéron et al., 1990). Clinical researches also indicate that consumption of 120 mg silybin twice a week for 2 months decreased aspartate transaminase and alanine transaminase levels in blood serum of patients with liver diseases (Wang et al., 1996).

One of the most effective property of silymarin is its ability to inhibit cellular permeability, which is associated with quantitative and qualitative alterations in membrane lipids. (Muriel et al., 1990) This suggests that silymarin may influence lipoprotein secretion and uptake. In this regard, silymarin and silibinin have been shown to reduce the turnover and synthesis of phospholipids in the rat liver. Moreover, silibinin has been shown to reduce labeled glycerol incorporation into lipids of isolated hepatocytes and also neutralize the inhibition of phospholipid synthesis (Videla et al., 1982)

Various studies have reported several mechanisms thorough which silymarin and its components act in liver diseases. These are summarized as, first It act as an antioxidant eliminating the reactive species and regulating the intracellular content of glutathione (Surai, 2015), Second it works as a cell membrane stabilizer and permeability regulator that prevents the entry of hepato toxic agents into hepatocytes (Trakulsrichai et al., 2017), third it works as a promoter of ribosomal RNA synthesis, which can stimulate hepatic regeneration (Bousserouel et al., 2012), next it works as an inhibitor of the transformation of stellate hepatocytes into myofbroblasts (Bode et al., 1977) and lastly it act as an anti-inflammatory drug as it suppresses the release of cytokines and inhibits the nuclear-kappa β transcription factor (NF-κB) (Federico et al., 2017). However, other studies have shown that the main mechanism that guarantees silymarin hepatoprotection is its ability to eliminate the reactive species (de Souza Santos et al., 2019).

3.2. Use in Breastfeeding

Breastfeeding is considered the optimal source of nutrition for infants from birth to 1 year and is supported by the American Academy of Pediatrics (AAP) and the World Health Organization. The 2012 AAP Policy on Breastfeeding recommends exclusive breastfeeding for 6 months, with continuation up to 1 year or longer. Breastfeeding has been associated with both short- and long-term benefits over formula feeding (Forinash et al., 2012). Breastfeeding for a minimum of 3 months decreased the risk of the infant developing acute otitis media and/or atopic dermatitis and if the child were exclusively breastfed for greater than 4 months, the result was a decrease in hospitalizations secondary to lower respiratory tract infection. Additionally, if breastfeeding continued for greater than 6 months, there was a potential decrease in the occurrence of acute lymphocytic leukemia and acute myeloid leukemia (Kent et al., 2012).Studies have also shown a potential decrease in death from sudden infant death syndrome and the development of asthma, diabetes mellitus, and obesity. Because of the many documented benefits of continued breastfeeding for the infant and mother, different measures are taken to increase breast milk production (Ip et al., 2007). A variety of herbal and pharmaceutical products have been recommended as galactogogues. One popular herbal galactogogue is milk thistle (silymarin, St. Mary's milk), which has been used for more than 2000 years for a variety of ailments. It is theorized that milk thistle galactogogue effects are secondary to an increase in prolactin levels, as seen in female rats (Capasso et al., 2009).

3.3. Anti-Oxidative Properties

Living organisms are constantly exposed to different toxins from the environment or as a result of biochemical reactions. Aerobic metabolism demands a constant oxygen flux for energy production, but the cost is the potential synthesis of reactive oxygen species (ROS) such as superoxide anion radical ($\bullet O_2^-$), hydroxyl radical ($\bullet OH^-$), and hydrogen peroxide (H_2O_2) (Halliwell, 2007). These highly reactive molecules possess the ability to bond with other cellular structures such as lipids, proteins, and DNA, increasing oxidative stress, causing damage. It is noteworthy that ROS are essential for maintaining homeostasis, contrary to it being thought that they are regarded as harmful (Schieber and Chandel, 2014). Therefore, it is crucial to maintain equilibrium between ROS production and elimination for homeostasis. In this regard, cells have developed a potent powerful system to protect themselves from damage. Nuclear factor erythroid 2 (NFE2)-related factor 2 (Nrf2) is known as a master regulator of antioxidant response; its activation is mediated by ROS (Bryan et al., 2013). The impaired Nrf2 pathway is associated with aging, inflammation, organ and tissue damage, and the consequent development of chronic diseases. The disruption of Nrf2 signaling has been associated with different pathological conditions (Vargas-Mendoza et al., 2019). It is reported that silymarin can work in diverse ways to reduce or prevent oxidative stress, acting directly on the scavenging of free radicals, blocking specific enzyme producers of free radicals, maintaining the integrity of the electron transport chain, contributing to optimal redox cell status, enhancing enzymes of the endogenous antioxidant systems, inducing non-enzymatic antioxidant defenses such as glutathione or transcription factors (Nrf2 and NF-κB), promoting signaling pathways of the redox response and, finally, activating the specific genes responsible for producing protector proteins such as sitruins, thioredoxins, and HSP (Surai, 2015). Further it has been have proven to participate in the activation of the Nrf2/Keap1/ARE pathway, regulating the expression of cytoprotective antioxidant enzymes such as NADPH and NQO1 (Vargas-Mendoza et al., 2020).

3.4. In Treatment of Neurological Disorders

Silymarin and its components are further used in treatment against neurological disorders such as Alzheimer's and Parkinson's disease (Devi et al., 2016). The flavonoid silibinin, which is isolated from silymarin, exhibited therapeutic activity against Alzheimer's disease by inducing an inhibitory mechanism on cholinesterase enzymes and on aggregation of amyloid beta (Aβ) peptide (Duan et al., 2015). Silymarin not only prevents the neurotoxicity caused by different disease pathologies but it also prevents the neurotoxicity induced by many chemicals like acrylamide. Treating PC12 cells with different concentrations of silymarin revealed that silymarin protects the cells by triggering Nrf2 signaling and Nrf2 expression (Li et al., 2017).Silymarin also prevents depression, which is a common neuropsychiatric disorder affecting the normal well-being of a person. When experimental mice were injected with 100 and 200 mg/kg of silymarin extract, it induced an antidepressant effect that mediated modulation of the corticosterone response (Thakare et al., 2016).

3.5. Anti-Ageing Properties

One of the main alterations associated with an increase in age is muscular atrophy and diminution in the functions of skeletal-muscle mobility and contractility (Gomes et al., 2017). Further, Diminution in muscular mass and its functionality have also been linked with down expression of Nrf2 (Hayes and Dinkova-Kostova, 2014). In wild-type and in Nrf2-/- young (4 months of age) and old (24 months of age) mice, a reduction of muscle mass and in contraction capacity has been observed in old mice with the Nrf2-/- genotype compared with those of the same age. The anti-aging effect of silymarin in the yeast model system revealed that silymarin increases the lifespan of animals by 10.1% and 24.8%, respectively. It was also able to reduce the progression of Alzheimer's disease in the transgenic CL4176 *C. elegans* model by reducing the expression of Aβ protein 1-42 in muscles (Kumar et al., 2015).

3.6. Anti-Diabetic Properties

A considerable number of herbs have been studied extensively regarding their hypoglycemic properties, which may be useful in the control of diabetes mellitus, one of the world fastest growing life style disease. Milk thistle is one of the herb which was regarded as a potent agent against insulin resistance and diabetes-induced hyperglycemia. Milk thistle has been shown to possess peroxisome proliferator-activated receptor γ (PPARγ) agonist properties. PPARγ is the molecular target of thiazolidinediones, which are used clinically as insulin sensitizers to lower blood glucose levels in diabetes type 2 patients. The thiazolidinedione type of PPARγ ligands is an agonist with a very high binding affinity (Kazazis et al., 2014). Further, oxidative stress is also reported as one of the main contributors to the pathogenesis of diabetes mellitus, it is believed that the antioxidant capacity of silymarin can improve several parameters in diabetic patients. In addition, the antiinflammatory, hypolipidemic and antihyperglycemic properties of silymarin may also have beneficial effects on diabetes and on different complications related to diabetes, such as nephropathy, neuropathy, retinopathy, metabolic disorders, liver disease and cardiomyopathy (Stolf et al., 2017).

3.7. Anti-Cancer Properties

In the last few decades, targeting cancer by the use of dietary phytochemicals has gained enormous attention. The reasonable reason and believe or mind set behind this fact is credited to either lesser or no side effects of natural compounds as compared to the modern chemotherapeutics, or due to their conventional use as dietary components by mankind for thousands of years

(Ram et al., 2020). Silymarin is a naturally derived polyphenol with amazingly high antioxidant and anti-tumorigenic properties. The anticancer properties of silymarin are determined by a variety of cellular pathways which include induction of apoptosis, cell cycle arrest, inhibition of angiogenesis and metastasis, induction of autophagy, induction of anti-oxidant response, modulation of the expression of aberrant miRNAs and epigenetic modulators, alteration in tumor microenvironment by modulating inflammatory response, and synergism with existing anti-cancer drugs (Tuli et al., 2021). Therefore, modulation of a vast array of cellular responses and homeostatic aspects makes silymarin an attractive chemotherapeutic agent.

Silymarin has been reported to be effective against number of carcinomas of the skin, liver, cervix, breast, lung, bladder, prostate, kidney and oral cancer (Ramasamy and Agarwal, 2008, Tuli et al., 2021). These studies have shown that silymarin could arrest the cell cycle at the G1/S phase through the induction of cyclin-dependent kinase inhibitors as p15, p21 and p27. In addition, it down-regulates the anti-apoptotic protein Bcl-2, up-regulates the pro-apoptotic protein Bax and inhibits the cell kinases (AKT, PKC and MAPK) and inflammatory transcription factors (e.g., NF-κB), among other effects involved in the proliferation and invasion of tumor cells in angiogenesis and metastasis. Furthermore, by the suppression of the STAT3 and ERK1/2 signaling pathways, it can inhibit oncogenesis, cell proliferation, cell migration and iNOS gene expression as well as induces caspase activation and apoptotic cell death in malignant cells (Agarwal et al., 2007).

4. Dietary Sources

Milk thistle plant itself is the major source of silymarin. All parts of the plant are reported to contain silymarin with the seeds containing high levels. Medical silymarin is extracted from milk thistle seeds (Tuli et al., 2021) and available in market in different formulations like as powder, capsules, supplements etc. The sprouts and young stems are eaten in many parts of the world-especially in Mediterranean countries. Silymarin is also present in most members of the thistle family but nowhere else in the high concentrations found in milk thistle. Besides milk thistle the most common food source of silymarin is artichokes (*Cynaras colymus*) which is also a member of the thistle family. Another rare member of the thistle family known as cardoon (*Cynara carduncului*) h is also known to contain silymarin. These plants are taken as common salad ingredient in Arabic countries.

Besides thistle family plants, some other spices like turmeric, coriander and plant like *Eclipta alba* is also reported of have small amounts of silymarin. The roots and rhizomes of the black cohosh (*Actaearace mosa*) also contain small amounts of silymarin. Beside these dark-skinned grapes, beet greens, peanuts, brewer's yeast and berries are also reported to have trace amounts of silymarin (Beedimani and Shetkar, 2015).

Conclusion

Silymarin is a valuable natural product that has many biological activities. Silymarin possess wide range of *in vitro* and *in vivo* mechanisms, such as antioxidant, anti-inflammatory, dose dependent anti-apoptotic and modifying cell transporters. Hence, it can be used as a promising medication in complementary medicine. In summary, silymarin is a very interesting chemical compound, the use of which as a dietary supplement is increasing all around the world, and it is evident from the recent studies aimed at increasing its oral bioavailability.

References

AbouZid, S., & Ahmed, O. M. (2013). Silymarin flavonolignans: structure–activity relationship and biosynthesis. In *Studies in Natural Products Chemistry* (Vol. 40, pp. 469-484). Elsevier.

Abrol, S., Trehan, A., & Katare, O. P. (2004). Formulation, characterization, and *in vitro* evaluation of silymarin-loaded lipid microspheres. *Drug Delivery*, 11, 185-191.

Agarwal, C., Tyagi, A., Kaur, M., & Agarwal, R. (2007). Silibinin inhibits constitutive activation of Stat3, and causes caspase activation and apoptotic death of human prostate carcinoma DU145 cells. *Carcinogenesis*, 28, 1463-1470.

Althagafy, H. S., Meza-Aviña, M. E., Oberlies, N. H., & Croatt, M. P. (2013). Mechanistic study of the biomimetic synthesis of flavonolign and its isomers in milk thistle. *The Journal of Organic Chemistry*, 78, 7594-7600.

Baer-Dubowska, W., Szaefer, H., & Krajka-Kuzniak, V. (1998). Inhibition of murine hepatic cytochrome P450 activities by natural and synthetic phenolic compounds. *Xenobiotica*, 28, 735-743.

Bahmani, M., Shirzad, H., Rafieian, S., & Rafieian-Kopaei, M. (2015). *Silybum marianum*: beyond hepatoprotection. *Journal of Evidence-Based Complementary & Alternative Medicine*, 20, 292-301.

Bai, Z. L., Tay, V., Guo, S. Z., Ren, J., & Shu, M. G. (2018). Silibinin induced human glioblastoma cell apoptosis concomitant with autophagy through simultaneous inhibition of mTOR and YAP. *BioMedResearch International, 2018*.

Beedimani, R. S., & Shetkar, S. (2015). Hepatoprotective activity of *Eclipta alba* against carbon tetrachloride-induced hepatotoxicity in albino rats. *International Journal of Basic & Clinical Pharmacology,* 4, 404-409.

Biedermann, D., Vavříková, E., Cvak, L., & Křen, V. (2014). Chemistry of silybin. *Natural Product Reports*, 31, 1138-1157.

Bijak, M. (2017). Silybin, a major bioactive component of milk thistle (*Silybum marianum* L. Gaernt.)-Chemistry, bioavailability, and metabolism. *Molecules*, 22, 1942.

Bode, J. C., Schmidt, U., & Dürr, H. K. (1977). Silymarin for the treatment of acute viral hepatitis? Report of a controlled trial (author's transl). *MedizinischeKlinik*, 72, 513-518.

Bousserouel, S., Bour, G., Kauntz, H., Gosse, F., Marescaux, J., & Raul, F. (2012). Silibinin inhibits tumor growth in a murine orthotopichepato carcinoma model and activates the TRAIL apoptotic signaling pathway. *Anticancer Research*, 32, 2455-2462.

Bryan, H. K., Olayanju, A., Goldring, C. E., & Park, B. K. (2013). The Nrf2 cell defense pathway: Keap1-dependent and-independent mechanisms of regulation. *Biochemical Pharmacology*, 85, 705-717.

Capasso, R., Aviello, G., Capasso, F., Savino, F., Izzo, A. A., Lembo, F., & Borrelli, F. (2009). Silymarin BIO-C®, an extract from *Silybum marianum* fruits, induces hyper prolactinemia in intact female rats. *Phytomedicine*, 16, 839-844.

de Souza Santos, V., Peters, B., Côco, L. Z., Alves, G. M., de Assis, A. L. E. M., Nogueira, B. V., Meyrelles, S. S., Porto, M. L., Vasquez, E. C., Campagnaro, B. P., & Pereira, T. M. C. (2019). Silymarin protects against radiocontrast-induced nephropathy in mice. *Life Sciences*, 228, 305-315.

Devi, K. P., Malar, D. S., Braidy, N., Nabavi, S. M., & Nabavi, S. F. (2017). A mini review on the chemistry and neuroprotective effects of silymarin. *Current Drug Targets*, 18, 1529-1536.

Di Sario, A., Bendia, E., Taffetani, S., Omenetti, A., Candelaresi, C., Marzioni, M., De Minicis, S., & Benedetti, A. (2005). Hepatoprotective and antifibrotic effect of a new silybin–phosphatidylcholine–vitamin E complex in rats. *Digestive and Liver Disease*, 37, 869-876.

Duan, S., Guan, X., Lin, R., Liu, X., Yan, Y., Lin, R., Zhang, T., Chen, X., Huang, J., Sun, X., Li, Q., Fang, S., Xu, J., Yao, J., & Gu, H. (2015). Silibinin inhibits acetylcholinesterase activity and amyloid β peptide aggregation: a dual-target drug for the treatment of Alzheimer's disease. *Neurobiology of Aging*, 36, 1792-1807.

European Pharmacopoeia Commission, & European Directorate for the Quality of Medicines & Healthcare. (2010). *European Pharmacopoeia* (Vol. 1). Council of Europe.

Federico, A., Dallio, M., & Loguercio, C. (2017). Silymarin/silybin and chronic liver disease: a marriage of many years. *Molecules*, 22, 191.

Flora, K., Hahn, M., Rosen, H., & Benner, K. (1998). Milk thistle (*Silybum marianum*) for the therapy of liver disease. *The American Journal of Gastroenterology*, 93, 139-143.

Forinash, A. B., Yancey, A. M., Barnes, K. N., & Myles, T. D. (2012). The use of galactogogues in the breastfeeding mother. *Annals of Pharmacotherapy*, 46, 1392-1404.

Foster, S. (1999). *Milk thistle: Silybum marianum*. American Botanical Council.

Gabay, M. P. (2002). Galactogogues: medications that induce lactation. *Journal of Human Lactation*, 18, 274-279.

Geier, J., Fuchs, T. H., & Wahl, R. (1990). Anaphylactic shock due to an extract of *Silybum marianum* in a patient with immediate-type allergy to kiwi fruit. *Allergologie*, 13, 387-388.

Gomes, M. J., Martinez, P. F., Pagan, L. U., Damatto, R. L., Cezar, M. D. M., Lima, A. R. R., Okoshi, K., & Okoshi, M. P. (2017). Skeletal muscle aging: influence of oxidative stress and physical exercise. *Oncotarget*, 8, 20428-20440.

Halliwell, B. (2007). Biochemistry of oxidative stress. *Biochemical Society Transactions*, 35, 1147-1150.

Hayes, J. D., & Dinkova-Kostova, A. T. (2014). The Nrf2 regulatory network provides an interface between redox and intermediary metabolism. *Trends in Biochemical Sciences*, 39, 199-218.

Ip, S., Chung, M., & Raman, G. (2007). Tufts-New England Medical Center Evidence-based Practice Center. Breastfeeding and maternal and infant health outcomes in developed countries. *Evidence Report/Technology Assessment*, 153, 1-186.

Karimi, G., Vahabzadeh, M., Lari, P., Rashedinia, M., & Moshiri, M. (2011). "Silymarin", a promising pharmacological agent for treatment of diseases. *Iranian Journal of Basic Medical Sciences*, 14, 308-317.

Kazazis, C. E., Evangelopoulos, A. A., Kollas, A., & Vallianou, N. G. (2014). The therapeutic potential of milk thistle in diabetes. *The Review of Diabetic Studies: RDS*, 11, 167.

Kent, J. C., Prime, D. K., & Garbin, C. P. (2012). Principles for maintaining or increasing breast milk production. *Journal of Obstetric, Gynecologic & Neonatal Nursing*, 41, 114-121.

Křen, V., & Walterová, D. (2005). Silybin and silymarin-new effects and applications. *Biomed Papers*, 149, 29-41.

Kroll, D. J., Shaw, H. S., & Oberlies, N. H. (2007). Milk thistle nomenclature: why it matters in cancer research and pharmacokinetic studies. *Integrative Cancer Therapies*, 6, 110-119.

Kumar, J., Park, K. C., Awasthi, A., & Prasad, B. (2015). Silymarin extends lifespan and reduces proteotoxicity in *C. elegans* Alzheimer's model. *CNS & Neurological Disorders-Drug Targets (Formerly Current Drug Targets-CNS & Neurological Disorders)*, 14, 295-302.

Kvasnička, F., Bıba, B., Ševčík, R., Voldřich, M., & Kratka, J. (2003). Analysis of the active components of silymarin. *Journal of Chromatography A*, 990, 239-245.

Ladas E, Kroll DJ, Kelly KA. Milk thistle (*Silybum marianum*). In: Coates PM, Blackman MR, Cragg GM, Levine M, Moss J, White JD, eds. *Encyclopedia of Dietary Supplements*. New York: Marcel Dekker; 2005:467-482.

Lee, D. Y. W., & Liu, Y. (2003). Molecular Structure and Stereochemistry of Silybin A, Silybin B, Isosilybin A, and Isosilybin B, Isolated from *Silybum marianum* (Milk Thistle). *Journal of Natural Products*, 66, 1171-1174.

Lettéron, P., Labbe, G., Degott, C., Berson, A., Fromenty, B., Delaforge, M., Larrey, D., &Pessayre, D. (1990). Mechanism for the protective effects of silymarin against carbon tetrachloride-induced lipid peroxidation and hepatotoxicity in mice: evidence that silymarin acts both as an inhibitor of metabolic activation and as a chain-breaking antioxidant. *Biochemical pharmacology*, 39, 2027-2034.

Li, L., Sun, H. Y., Liu, W., Zhao, H. Y., & Shao, M. L. (2017). Silymarin protects against acrylamide-induced neurotoxicity via Nrf2 signalling in PC12 cells. *Food and Chemical Toxicology*, 102, 93-101.

Maheshwari, H., Agarwal, R., Patil, C., & Katare, O. P. (2003). Preparation and pharmacological evaluation of silibinin liposomes. *Arzneimittelforschung*, 53, 420-427.

Maryana, W., Rachmawati, H., &Mudhakir, D. (2016). Formation of phytosome containing silymarin using thin layer-hydration technique aimed for oral delivery. *Materials Today: Proceedings*, 3, 855-866.

Mastron, J. K., Siveen, K. S., Sethi, G., &Bishayee, A. (2015). Silymarin and hepatocellular carcinoma: a systematic, comprehensive, and critical review. *Anti-Cancer Drugs*, 26, 475-486.

Mayer, K. E., Myers, R. P., & Lee, S. S. (2005). Silymarin treatment of viral hepatitis: a systematic review. *Journal of Viral Hepatitis*, 12, 559-567.

Murata, N., Murakami, K., Ozawa, Y., Kinoshita, N., Irie, K., Shirasawa, T., & Shimizu, T. (2010). Silymarin attenuated the amyloid β plaque burden and improved behavioral abnormalities in an Alzheimer's disease mouse model. *Bioscience, Biotechnology, and Biochemistry*, 74, 2299-2306.

Muriel, P., & Mourelle, M. (1990). The role of membrane composition in ATPase activities of cirrhotic rat liver: effect of silymarin. *Journal of Applied Toxicology*, 10, 281-284.

Ram, G., Sharma, V. R., Sheikh, I., Sankhyan, A., Aggarwal, D., & Sharma, A. K. (2020). Anti-cancer potential of natural products: recent trends, scope and relevance. *Letters in Applied NanoBioScience*, 9, 902-907.

Ramasamy, K., & Agarwal, R. (2008). Multitargeted therapy of cancer by silymarin. *Cancer Letters*, 269, 352-362.

Schieber, M., & Chandel, N. S. (2014). ROS function in redox signaling and oxidative stress. *Current Biology*, 24, R453-R462.

Sheikh, I., Sharma, V., Tuli, H. S., Aggarwal, D., Sankhyan, A., Vyas, P., Sharma, A. K., &Bishayee, A. (2021). Cancer chemoprevention by flavonoids, dietary polyphenols and terpenoids. *Biointerface Research in Applied Chemistry*, 11, 8502-8537.

Si, L., Liu, W., Hayashi, T., Ji, Y., Fu, J., Nie, Y., Mizuno, K., Hattori, S., Onodera, S., & Ikejima, T. (2019). Silibinin-induced apoptosis of breast cancer cells involves mitochondrial impairment. *Archives of Biochemistry and Biophysics*, 671, 42-51.

Šimanek, V. I. L. Í. M., Kren, V., Ulrichová, J., Vicar, J., & Cvak, L. (2000). Silymarin: what is in the name? An appeal for a change of editorial policy. *Hepatology*, 32, 442-444.

Song, Z., Deaciuc, I., Song, M., Lee, D. Y. W., Liu, Y., Ji, X., & McClain, C. (2006). Silymarin protects against acute ethanol-induced hepatotoxicity in mice. *Alcoholism: Clinical and Experimental Research*, 30, 407-413.

Sonnenbichler, J., & Zetl, I. (1986). Biochemical effects of the flavonolignone silibinin on RNA, protein and DNA synthesis in rat livers. *Progress in Clinical and Biological Research*, 213, 319-331.

Stolf, A. M., Cardoso, C. C., & Acco, A. (2017). Effects of silymarin on diabetes mellitus complications: a review. *Phytotherapy Research*, 31, 366-374.

Surai, P. F. (2015). Silymarin as a natural antioxidant: an overview of the current evidence and perspectives. *Antioxidants*, 4, 204-247.

Thakare, V. N., Dhakane, V. D., & Patel, B. M. (2016). Potential antidepressant-like activity of silymarin in the acute restraint stress in mice: Modulation of corticosterone and oxidative stress response in cerebral cortex and hippocampus. *Pharmacological Reports*, 68, 1020-1027.

Trakulsrichai, S., Sriapha, C., Tongpoo, A., Udomsubpayakul, U., Wongvisavakorn, S., Srisuma, S., &Wananukul, W. (2017). Clinical characteristics and outcome of toxicity from Amanita mushroom poisoning. *International Journal of General Medicine*, 10, 395-400.

Tuli, H. S., Mittal, S., Aggarwal, D., Parashar, G., Parashar, N. C., Upadhyay, S. K., Barwa, T. S., Jain, A., Kaur, G., Savla, R., Sak, K., Kumar, M., Varol, M., Iqubal, A., & Sharma, A. K. (2021). Path of Silibinin from diet to medicine: A dietary polyphenolic flavonoid having potential anti-cancer therapeutic significance. *Seminars in Cancer Biology*, 73, 196-218). https://doi.org/10.1016/j.semcancer. 2020.09.014.

Vargas-Mendoza, N., Morales-González, Á., Madrigal-Santillán, E. O., Madrigal-Bujaidar, E., Álvarez-González, I., García-Melo, L. F., Anguiano-Robledo, L., Fregoso-Aguilar, T., & Morales-Gonzalez, J. A. (2019). Antioxidant and adaptative response mediated by Nrf2 during physical exercise. *Antioxidants*, 8, 196.

Vargas-Mendoza, N., Morales-González, Á., Morales-Martínez, M., Soriano-Ursúa, M. A., Delgado-Olivares, L., Sandoval-Gallegos, E. M., Madrigal-Bujaidar, E., Álvarez-González, I., Madrigal-Santillán, E., & Morales-Gonzalez, J. A. (2020). Flavolignans from Silymarin as Nrf2 bioactivators and their therapeutic applications. *Biomedicines*, 8, 122.

Videla, L. A., & Valenzuela, A. (1982). Alcohol ingestion, liver glutathione and lipoperoxidation: metabolic interrelations and pathological implications. *Life Sciences*, 31, 2395-2407.

Wagner, H., Hörhammer, L., & Münster, R. (1968). On the chemistry of silymarin (silybin), the active principle of the fruits from *Silybum marianum* (L.) Gaertn. (*Carduus marianus* L.). *Arzneimittel-Forschung*, 18, 688-696.

Wang, M., La Grange, L., Tao, J., & Reyes, E. (1996). Hepatoprotective properties of *Silybum marianum* herbal preparation on ethanol-induced liver damage. *Fitoterapia (Milano)*, 67, 166-171.

Wesołowska, O., Łania-Pietrzak, B., Kużdżał, M., Stańczak, K., Mosiądz, D., Dobryszycki, P., Ożyhar, A., Komorowska, M., Hendrich, A. B., &Michalak, K. (2007). Influence of silybin on biophysical properties of phospholipid bilayers 1. *Acta Pharmacologica Sinica*, 28, 296-306.

Woodward, M. (1994). Gerard's herbal: The history of plants. *London: Studio Editions, Ltd*.

Yang, Z., Zhuang, L., Lu, Y., Xu, Q., & Chen, X. (2014). Effects and tolerance of silymarin (milk thistle) in chronic hepatitis C virus infection patients: a meta-analysis of randomized controlled trials. *Bio Med Research International*, 941085.https://doi.org/10.1155/2014/941085

Chapter 5

Biological Properties and Therapeutic Applications of Luteolin: Current Status and Future Perspectives

Kumar Preet[1], Nirmala Sehrawat[1,*], Anil Kumar Sharma[1], Sachin Kumar[2], Amandeep Singh[1], Rachna Nara[1], Nipunjot Kaur[3] and Vikas Kumar[1]

[1]Department of Biotechnology, Maharishi Markandeshwar (Deemed to be University), Mullana-Ambala, India

[2]Department of Bioinformatics, Janta Vedic College, Baraut- Baghpat, Uttar Pradesh, India

[3]G.S.S.D.G.S. Khalsa College, Patiala, Punjab, India

Abstract

Flavonoids are well known for their health benefits. The flavonoids are considered as an indispensable component in a variety of nutraceutical, pharmaceutical, medicinal and cosmetic applications. This is attributed to their anti-microbial, anti-oxidative, anti-inflammatory, anti-mutagenic and anti-carcinogenic properties coupled with their capacity to modulate cellular and metabolic pathways. Flavonoids are now well known for their biological properties and health benefits. Luteolin is one of the well known flavonoids abundantly present in plants based diet. It has been reported for its anti-cancer, anti-oxidant, anti-apoptotic, and anti-inflammatory properties. Luteolin has been reported for the protective nature against viruses also. It may be a promising molecule of interest

* Corresponding Author's Email: nirmalasehrawat@gmail.com.

In: Flavonoids
Editors: Nirmala Sehrawat, Diwakar Aggarwal, Sushil Kumar Upadhyay et al.
ISBN: 978-1-68507-953-6

against COVID-19 disease. Dietary intake and protective role of this flavonoid requires elaborated and extensive research for effective management of various diseases and viral infection.

Keywords: luteolin, flavonoid, biological properties, COVID-19

1. Introduction

Flavonoids are the natural substances having variable phenolic structures. These are well known to provide various health benefits. Flavonoids are mainly present in fruits, vegetables, grains, bark, roots, stems, flowers, tea and processed food based on plant materials (Middleton, 1998). These natural products are well known for their beneficial effects on health.

Flavonoids can be divided into various classes on the basis of their molecular structure (Rice-Evans et al., 1996). The flavones are characterized by a planar structure because of a double bond in the central aromatic ring. One of the best described flavonoids, Luteolin, is a member of this group. It is found in significant amount in celery seeds, oregano, parsley, bird chili, juniper berries & onion leaves. The second group is the flavanones, which are mainly found in citrus fruit. An example of a flavonoid of this group is naringin. Flavonoids belonging to the catechins are mainly found in green and black tea and in red wine (de Groot and Rauen 1998), whereas anthocyanins are found in strawberries and other berries, grapes, wine, and tea.

Luteolin is a polyphenolic flavone found in many herbs, fruits and vegetables and traditionally used in Chinese medicine, which has been found to have anti-cancer, anti-oxidant, anti-apoptotic, and anti-inflammatory properties (Molnar et al., 1981; Selvi et al., 2015; Xia et al., 2016). Luteolin was reported to have protective and anti-inflammatory effects in models of traumatic brain injury, spinal cord injury, and MPP+ induced toxicity (Wruck et al., 2007; Paterniti et al., 2013; Xu et al., 2014). Belonging to the flavone group of flavonoids, luteolin has a C6-C3-C6 structure and possesses two benzene rings (A, B), a third, oxygen-containing (C) ring, and a 2–3 carbon double bond. Luteolin also possesses hydroxyl groups at carbons 5, 7, 3', and 4' positions (Ross and Kasum 2002). The hydroxyl moieties and 2–3 double bond are important structure features in luteolin that are associated with its biochemical and biological activities (Chan et al., 2003). As in other flavonoids, luteolin is often glycosylated in plants, and the glycoside is hydrolyzed to free luteolin during absorption (Hempel et al., 1999). Some portion of luteolin is converted to glucuronides when passing through the intestinal mucosa (Shimoi et al., 1998). Luteolin is heat stable and losses due to cooking are relatively low (Le Marchand, 2002). Various health benefits and therapeutic properties of luteolin have been reported. It has been found effective in management of various diseases also. Extensive research is required to further expand the knowledge about the therapeutic potential of luteolin and its use in management of diseases. This chapter focuses the dietary sources of this flavonoid along with biological properties and therapeutic potential.

2. Dietary Sources

Luteolin is a naturally occurring plant-derived flavonoid that is widely distributed in edible herbs, vegetable and fruits such as bird chilli, oregano, juniper berries, celery seeds and parsley in significant amount (Miean and Mohamed 2001). Luteolin can be found in Bryophyta, Pteridophyta, Pinophyta, and Magnoliophyta, among other botanical groups (Boopathi et al., 2020). Lutein glycosides have been discovered in mosses belonging to the Hedwigiaceae family in the Bryophyta. Many edible plants have been found to contain luteolin (Pandurangan and Esa, 2014). Some of the luteolin containing dietary sources include carrots (*Daucus carota*), peppers (*Capsicum annuum*), celery (*Apium graveolens*), olive oil (*Olea europaea*), peppermint (*Menthaxpiperita*), pomegranate (*Punica granatum*) and cucumber (*Cucumis sativus*). Luteolin is abundantly present in various fruits and vegetables such as parsley, thyme, celery, olive oil, onion leaves, broccoli, peppermint and peppers. It is also present in rosemary, oregano, cabbages, carrots, artichokes, chrysanthemum flowers, apple skins and other fruits or vegetables in significant amount.

3. Biological Properties and Therapeutic Potential of Flavonoids

Luteolin has been reported for various biological properties (Figure 1) and therapeutic potential. Some of the properties have been discussed here.

3.1. Antiviral Potential of Flavonoid (Luteolin) Against SARS Viruses and COVID-19

Flavonoids are a class of phenolic natural products, well-identified in traditional and modern medicines in the management of several diseases including viral infections. Flavonoids have been shown to have potential inhibitory activity against various viruses including coronaviruses (Alzaabi et al., 2021; Jo et al., 2020; Russo et al., 2021). Several flavonoids including luteolin have been found effective against corona virus. Flavonoids can demonstrate potential inhibitory activity against SARS-CoV-2 by binding to essential viral targets required in virus entry and/ or replication (Alzaabi et al., 2021).

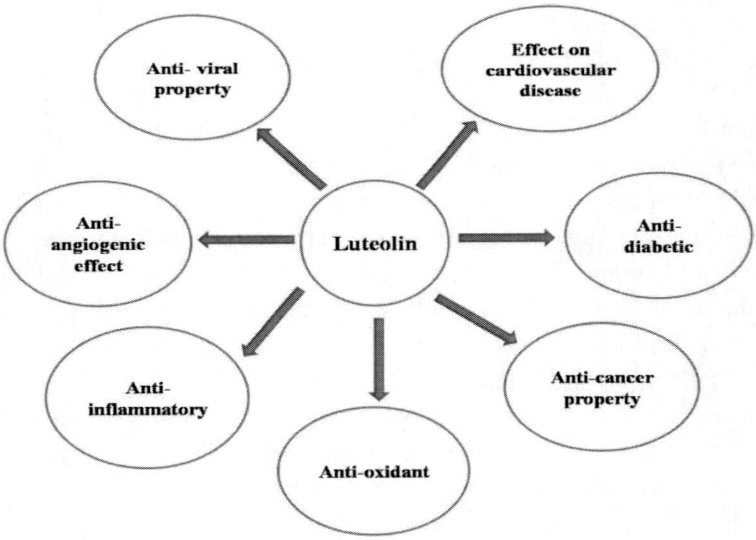

Figure 1. Major biological properties and therapeutic effects of flavonoid luteolin.

A recent report highlighted that flavonoids in particular; quercetin and luteolin can exhibit promising multi-target activity against SARS-CoV-2 (Alzaabi et al., 2021). SARS-CoV-2 can develop severe complications including septic shock and multiple organ failure that may result in death especially in people at high risk including immuno-compromised and those with underlying medical conditions such as cancers, diabetes, cardio-vascular disease, and chronic respiratory diseases. A computational study indicated the potential importance of luteolin-7-glucoside, as potential inhibitors to SARS-

CoV-2 Mpro (Khaerunnisa et al., 2020). Luteolin is one such compound which is used as an anti-inflammatory agent. Several studies have explained the protective nature of luteolin by inhibiting virus entry and fusion with human receptors in old SARS-CoV that had emerged in 2003. Thus, regular consumption of foods having adequate amount of luteolin in our diet may be helpful in inhibiting the SARS-CoV-2 infection as well and may prevent the consequent symptoms in COVID-19 patients.

Recently, Ansari et al., (2020) has reported the dietary flavonoid luteolin as potential anti-COVID-19 agent. According to authors, luteolin is among the important plant based flavonoids that is widely distributed in edible herbs and vegetables. The pandemic coronavirus disease-2019 (COVID-19) is a disease of the respiratory tract that consists of mild to severe symptoms of pneumonia including fever, muscle aches, sore throat, coughing, and shortness of breath. The situation has prompted the researchers to evaluate the bioactive compounds that can be used to prevent respiratory-related illness. Luteolin is dietary compound which is used as an anti-inflammatory agent. Several studies have explained the protective nature of luteolin by inhibiting virus entry and fusion with human receptors in old SARS-CoV that had emerged in 2003 (Ansari et al., 2020). Therefore, regular consumption of foods having adequate amount of luteolin in our diet may be helpful in inhibiting the SARS-CoV-2 infection. Also, it may prevent the consequent symptoms in COVID-19 patients. Ansari et al. (2020) investigated the molecular docking studies of luteolin with six different SARS-CoV-2 encoded key proteins. It has been found that luteolin showed excellent inhibitory action against papain-like proteinase which is a main protease of SARS-CoV-2. Luteolin was also found several times more active than remdesivir. Therefore, authors recommended that the foods rich in luteoline in significant amount might be explored further for potential use against COVID-19 (Ansari et al., 2020).

This area requires extensive research and clinical trials to establish the effectiveness of luteolin against the COVID-19.

3.2. Anti-Inflammatory Property

Inflammation is one of the important defence mechanisms of the body. In inflammation, macrophages are activated by various molecules, including cytokines from the host and toxins from the pathogens. The activated macrophages produce various inflammatory molecules including tumor necrosis factor α (TNFα), interleukins (ILs), and free radicals (ROS and reactive nitrogen species, RNS), leading to recruitment of inflammatory cells, such as neutrophils and lymphocytes, to the infection site and clearance of the pathogens (Brody and Spira, 2006; Karin et al., 2006; Lin et al., 2008). Luteolin has been reported for its anti-inflammatory effect. Luteolin shows its anti-inflammatory effect through suppressing the production of specific cytokines and their signal transduction pathways. Luteolin has been found to inhibit LPS-stimulated TNFα and IL-6 release which further regulates LPS-induced activation of nuclear factor kappa B (NF-κB) and mitogen-activated protein kinase (MAPK) family members ERK, p38, and JNK (Lin et al., 2008). In neurodegenerative illnesses and neuronal cell death, luteolin inhibits inflammation in brain tissues while also regulating various cell signaling pathways (Park and Song, 2013; Nabavi et al., 2015).

3.3. Anti-Cancer Potential of Luteolin

Luteolin has been reported to have anti-caner properties. Luteolin's anticancer property is associated with the induction of apoptosis, and inhibition of cell proliferation, metastasis and angiogenesis. Luteolin has also been found to inhibit the metabolism of carcinogens that generates active mutagens in liver microsomes (Lin et al., 2008). Luteolin is able to arrest the cell cycle during the G1 phase in human gastric and prostate cancer, and in melanoma cells (Matsukawa et al., 1993; Kobayashi et al., 2002; Casagrande et al., 2001; Lin et al., 2008). Luteolin has been found to kills cancer cells by inducing apoptotic cell death in many types of cancer cells. The preventive and protective role of luteolin has been investigated against various cancer cells including breast cancer cells, prostate cancer cells, CRC, epidermoid carcinoma, leukemia, pancreatic tumor, and hepatoma. The luteolin has been found promising and molecule of interest against specific cancers. Further evidence based research is required for establishing its possible use against cancer.

3.4. Preventive Role in Cardiovascular Disease (CVD)

Cardio-protective role of luteolin has been reported. It has been found to improve the heart function by enhancing contractility, up-regulating autophagy, and preventing cardiac fibrosis. Further, the anti-atherosclerosis properties of luteolin stem from inhibiting the proliferation and migration of

VSMCs, decreasing inflammation and ameliorating oxidative damage (Luo et al., 2017). Research on mechanistic insights has been reported for the luteolin based cardiac protection. Further research on various aspects including mechanisms, dosage, clinical trails and related aspects are required.

3.5. Anti-Diabetic Property of Luteolin

Diabetes is very common through out the world. In diabetes, the pancreas becomes dysfunctional and become unable to secrete enough insulin for glucose metabolism due to destruction of β-cell (Ambasta et al., 2019). Luteolin can promote the regeneration of β-cell and hence higher insulin secretion for glucose metabolism. Luteolin mode of action is via lowering ROS and promoting the cell proliferation in pancreas or inhibiting apoptosis in the pancreas (Ambasta et al., 2019). The mechanism of regeneration of the pancreas by luteolin is by diverse signaling molecules. Luteolin has also been reported for its anti-diabetic property because of its anti-oxidant potential. Oxidative stress is a prerequisite for the pathogenesis of diabetes. Luteolin has the protective effect on diabetic cardiomyopathy, diabetic nephropathy and oxidative stress in diabetic condition (Qian et al., 2010; Wang et al., 2011; 2012; Ambasta et al., 2019). The levels of free radical and antioxidant in the cell determine the initiation and progression of disease. Luteolin down-regulates the ROS levels by up-regulating SOD and cytokine levels. SOD is known for its antioxidant potential and hence lowers the overall ROS level for radical homeostasis (Ambasta et al., 2019).

3.6. Antioxidant and Antimicrobial Activity of Luteolin

Luteolin is common flavonol glycoside compounds found in many plants that possess multiple biological activities. One of the important biological activities is its antioxidant property. It has been found that luteolin possess significant antioxidant activity. Luteolin has been found to inhibit ROS-induced damage of lipids, DNA, and protein (Robak et al., 1988; Brown and Rice-Evans, 1998; Lin et al., 2008). Various other authors have also reported the antioxidant activity of luteolin, the exact mechanism involved in humans should be elaborated and studied extensively. Besides antioxidant activity, luteolin have been shown to have anti-microbial activity against selected microbial strains.

Conclusion and Future Perspectives

The flavonoid luteolin is distributed in various plants and plant parts that can be included in human diet. It has been found to have anti-oxidant, antimicrobial and anti-inflammatory properties. Also, it has been reported for its therapeutic properties including anti-diabetic, anti-cancerous, effective against cardiovascular disease, anti-viral properties and various other properties. The flavonoid luteolin can be effective in management of various diseases and it can provide various health benefits to the consumer. Luteolin has the potential to be used as safe and alternative treatment for some human diseases. More extensive and elaborated research including clinical studies is required on this flavonoid for effective usage in managing the human diseases.

Acknowledgment

Authors are thankful to Head, Department of Biotechnology, Maharishi Markandeshwar (Deemed to be University) for necessary help and support.

References

Alzaabi M.M., Hamdy R., Ashmawy N.S., Hamoda A.M., Alkhayat F., Khademi N.N., Al Joud S.M.A., El-Keblawy A.A., Soliman S.S.M. (2021). Flavonoids are promising safe therapy against COVID-19. *Phytochem Rev.* 1-22. doi: 10.1007/s11101-021-09759-z.

Ambasta R.K., Gupta R., Kumar D., Bhattacharya S., Sarkar A., Kumar P. (2019). *Can luteolin be a therapeutic molecule for both colon cancer and diabetes? Briefings in Functional Genomics*, 18(4), 230–239.

Ansari W.A., Ahamad T., Khan M.A., Khan Z.A., Khan M.F. (2020). Luteolin: A dietary molecule as potential anti- COVID-19 agent. *Research Square*. doi: 10.21203/rs.3.rs-35368/v1.

Boopathi S., Poma A.B., Kolandaivel P. (2020) Novel 2019 coronavirus structure, mechanism of action, antiviral drug promises and rule out against its treatment. *J. Biomol. Struct. Dyn.*,1–10. doi: 10.1080/07391102.2020.1758788.

Brody J.S., Spira A. (2006). State of the art. chronic obstructive pulmonarydisease, inflammation, and lung cancer. *Proc. Am.Thorac. Soc.*, 3, 535–537.

Brown J.E., Rice-Evans C.A. (1998). Luteolin-rich artichoke extract protects low density lipoprotein from oxidation in vitro. *Free Radic. Res.* 29, 247–255.

Casagrande F., Darbon J.M. (2001). Effects of structurally related flavonoids on cell cycle progression of human melanoma cells: regulation of cyclin-dependent kinases CDK2 and *CDK1. Biochem. Pharmacol.* 61, 1205–1215.

Chan T.S., Galati G., Pannala A.S., Rice-Evans C., O'Brien P.J. (2003). Simultaneous detection of the antioxidant and pro-oxidant activity of dietary polyphenolics in a peroxidase system. *Free Radic. Res.* 37(7), 787-794.

deGroot H., Rauen U. (1998). Tissue injury by reactive oxygen species and the protective effects of flavonoids. *Fundam. Clin. Pharmacol.* 12(3), 249-255.

Hempel J., Pforte H., Raab B., Engst W., Bohm H., Jacobasch G. (1999). Flavonols and flavones of parsley cell suspension culture change the antioxidative capacity of plasma *in rats. Nahrung,* 43(3), 201-204.

Jo S., Kim S., Kim D.Y., Kim M.S., Shin D.H. (2020). Flavonoids with inhibitory activity against SARS-CoV-2 3CLpro. *J. Enzyme Inhib. Med.Chem.* 35(1), 1539-1544. doi: S10.1080/14756366.2020.1801672.

Karin M., Lawrence T., Nizet V. (2006). Innate immunity gone awry: linking microbial infections to chronic inflammation and cancer. *Cell* 124, 823–835.

Khaerunnisa S., Kurniawan H., Awaluddin R., Suhartati S., Soetjipto S. (2020). Potential inhibitor of COVID-19 main protease (Mpro) from several medicinal plant compounds by molecular docking study. doi: 10.20944/preprints202003.0226.v1.

Kobayashi T., Nakata T., Kuzumaki T. (2002). Effect of flavonoids on cell cycle progression in prostate cancer cells. *Cancer Lett.,* 176, 17–23.

Le Marchand L. (2002). Cancer preventiveeffects of flavonoids--a review. *Biomed. Pharmacother*, 56(6), 296-301.

Lin Y., Shi R., Wang X., Shen H. (2008). Luteolin, a flavonoid with potentials for cancer prevention and *therapy. Curr. Cancer Drug Targets*, 8(7), 634–646.

Luo Y., Shang P., Li D. (2017) Luteolin: A flavonoid that has multiple cardio-protective effects and its molecular mechanisms. *Front Pharmacol.,* 8, 692. doi: 10.3389/fphar.2017.00692.

Luo Y., Shang P., Li D. (2017). Luteolin: A flavonoid that has multiple cardio-protective effects and its molecular mechanisms. *Front Pharmacol.*, 8, 692. doi: 10.3389/fphar.2017.00692.

Matsukawa Y., Marui N., Sakai T., Satomi Y., Yoshida M., Matsumoto K., Nishino H., Aoike A. (1993). Genistein arrests cell cycle progression at G2-M. *Cancer Res.*, 53, 1328–1331.

Middleton E.J. (1998). Effect of plant flavonoids on immune and inflammatory cell function. *Adv.Exp. Med.Biol.,* 439, 175-182.

Miean K.H., Mohamed S. (2001). Flavonoid (myricetin, quercetin, kaempferol, luteolin, and apigenin) content of edible tropical plants. *J.Agric. Food Chem.,* 49(6), 3106-3112.

Molnár J., Béládi I., Domonkos K., Földeák S., Boda K., Veckenstedt A. (1981). Antitumoractivity of flavonoids on NK/Ly ascites tumorcells. *Neoplasma*, 28(1), 11-18.

Nabavi S.F., Braidy N., Gortzi O., et al. (2015). Luteolin as an anti-inflammatory and neuroprotective agent: a brief review. *Brain Res. Bull.*, 119, 1–11.

Pandurangan A. K., Norhaizan M. Esa. (2014). Luteolin, a bioflavonoid inhibits colorectal cancer through modulation of multiple signaling pathways: a review. *Asian Pacific Journal of Cancer Prevention*. 15, 5501-5508.

Park C.M., Song Y.S. (2013). Luteolin and luteolin-7-O-glucoside inhibit lipopolysaccharide-induced inflammatory responses through modulation of NF-kappaB/AP-1/PI3K-Akt signaling cascades in RAW 264.7 cells. *Nutr. Res. Pract.,* 7, 423–429.

Paterniti I., Impellizzeri D., Di Paola R., Navarra M., Cuzzocrea S., Esposito E. (2013). A new co-ultramicronized composite including palmitoylethanolamide and luteolin to prevent neuroinflammation in spinal cordinjury. *J. Neuroinflammation*, 10, 91.

Qian L.B., Wang H.P., Chen Y., et al. (2010). Luteolinreduces high glucose-mediated impairment of endothelium-dependent relaxation in rat aorta by reducing oxidative stress. *Pharmacol. Res.* 61(4), 281–287.

Rice-Evans C.A., Miller N.J., Paganga G. (1996). Structure-antioxidant activity relationships of flavonoids and phenolicacids. *Free Radic.Biol. Med.,* 20(7), 933-956.

Robak J., Shridi F., Wolbis M., Krolikowska M. (1988). Screening of the influence of flavonoids on lipoxygenase and cyclooxygenase activity, as well as on nonenzymic lipid oxidation. *Pol. J.Pharmacol. Pharm*., 40, 451–458.

Ross J.A., Kasum C.M. (2002). Dietary flavonoids: bioavailability, metaboliceffects, and safety. *Annu.Rev.Nutr.,* 22, 19-34.

Russo M., Moccia S., Spagnuolo C., Tedesco I., Russo GL. (2020). Roles of flavonoids against coronavirus infection. *Chem. Biol. Interact.* 328, 109211. doi: 10.1016/j.cbi. 2020.109211.

Selvi R.B., Swaminathan, A., Chatterjee, S., Shanmugam, M.K., Li F., Ramakrishnan G.B. (2015). Inhibition of p300 lysine acetyltransferase activity by luteolin reduces tumor growth in head and neck squamous cell carcinoma (HNSCC) xenograft mouse model. *Oncotarget*, 6(41), 43806-43818.

Wang G., Li W., Lu X., et al. (2012). Luteolin ameliorates cardiac failure in type I diabetic cardiomyopathy. *J. Diabetes Complications* 26(4), 259–265.

Wang G.G., Lu X.H., Li W., et al. (2011). Protective effects of luteolin on diabetic nephropathy in STZ-induced diabetic rats. *Evid. Based Complement Alternat. Med*., 323171.

Wruck C.J., Claussen M., Fuhrmann G., Römer L., Schulz A., Pufe T. (2007). Luteolin protects rat PC12 and C6 cells against MPP+ induced toxicity via an ERK dependent Keap1-Nrf2-ARE pathway. *J. Neural. Transm. Suppl*, (72), 57-67.

Xia N., Chen G., Liu M., Ye X., Pan Y., Ge J. (2016). Anti-inflammatory effects of luteolin on experimental autoimmune thyroiditis in mice. *Exp Ther Med*, 12(6), 4049-4054.

Xu J., Wang H., Ding K., Zhang L., Wang C., Li T. (2014). Luteolin provides neuroprotection in models of traumatic brain injury via the Nrf2-ARE pathway. *Free Radic Biol Med*, 71, 186-195.

Chapter 6

Therapeutic Potential of Epigallocatechin: Current Status and Future Perspectives

Sneh Lata[1,*], Mukesh Yadav[2], Nirmala Sehrawat[2], Kumar Preet[2] and Manoj Singh[2]

[1]University Institute of Biotechnology, Chandigarh University, Gharuan, Punjab, India

[2]Department of Biotechnology, Maharishi Markandeshwar (Deemed to be University), Mullana-Ambala, Haryana, India

Abstract

People are focusing on flavonoid-rich diets to obtain maximal health benefits. Flavonoids are well known to provide various health benefits. Also, they are important in the prevention and management of different diseases. Epigallocatechin is one of the important flavonoids known for its health benefits. Currently, green tea is of great attraction due to the presence of flavonoids and catechins. The catechins present in these teas have grabbed a lot of attention due to their antimicrobial, anti-inflammatory, anticancer, anti-obesity, anti-glycation, as well as neuro- and cardioprotective effects. The present chapter focuses on the potential of epigallocatechin as an antioxidant, antimicrobial, anticancer, anti-inflammatory agent, anti-aging agent as well as its role in preventing cardiovascular diseases.

Keywords: catechins, epigallocatechin, antibacterial, anti-cancer, cardiovascular

[*] Corresponding Author's Email: sneh.uibt@cumail.in.

In: Flavonoids
Editors: Nirmala Sehrawat, Diwakar Aggarwal, Sushil Kumar Upadhyay et al.
ISBN: 978-1-68507-953-6

1. Introduction

Tea is the most commonly consumed beverage in the world. It has a very delicate and aromatic flavor in it which also corresponds to its health benefits. Polyphenols played an important role in preventing human diseases (Casas et al., 2018; Fraga et al., 2019; Rodriguez et al., 2019;Bungau et al., 2019; Poti et al., 2019). Flavonoids have great attention among polyphenols which are mainly found in fruits, vegetables, green tea, red wine, etc. (Matute et al., 2020).Among all these sources, green tea has been massively consumed in Asian countries for eras as a liquid refreshment (Younes et al., 2018). It's a very popular drink as it grabbed my attention as weight loss drink. The source of green tea is the plant *Camellia sinensis* which grows best in tropical and subtropical regions (Gupta et al., 2014) and is studied for numerous health benefits (Reygaert, 2014). The health benefits of green tea are due to the presence of catechins in them which makes them attractive for researchers. Tea catechins include(−)-epigallocatechin (EGC), (−)-epicatechin (EC), (−)-epicatechin-3-gallate (ECG), and (−)-epigallo-catechin-3-gallate (EGCG) (Jin et al., 2006), which are predominantly present in green tea (Fung et al., 2012).

Epigallocatechin (EGC) is one of the most common catechins which are found in green as well as black tea. It is a colorless bioactive (Bradfield et al., 1947; Bradfield et al., 1950) monomeric flavonoid with a molecular weight of 306.4 Da. Its persistence in black tea is 5-91 mg/l. EGC is making up approximately 19% of green tea. (Cabrera et al., 2006; Jigisha et al., 2012; McKay and Blumberg, 2002). The amount of EGC in a tea depends on its origin. For instance, 1 gram of Japanese dry tea has 20mg of EGC (Vuong et al., 2011) in it. Epigallocatechin (EGC) is hydrophilic with a three-ring structure with six hydroxyl groups (Ambigaipalan et al., 2020)as shown in Figure 1. The persistence of epigallocatechin in green tea led to various health-promoting effects (Vergote et al., 2002; Yang et al., 1998).

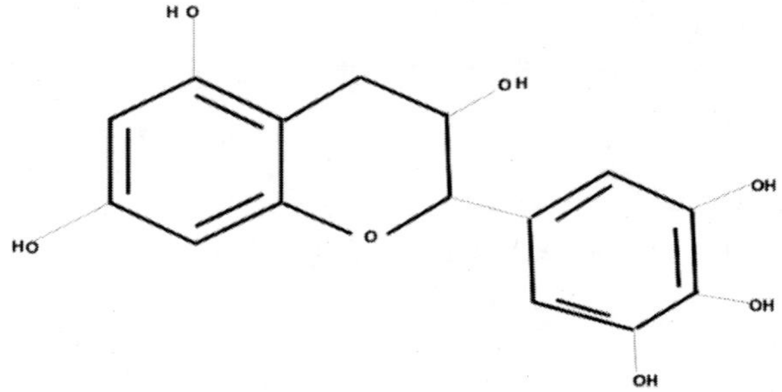

Figure 1. Structure of epigallocatechin.

2. Dietary and Natural Resources of Epigallocatechin

The most attention-grabbing source of epigallocatechin (EGC) is the *Camellia sinensis* plant from which almost all teas including black, green, oolong as well as white tea derives. EGC is found in apples, raspberries, black grapes, blackberries, red wine, apricots, broad beans, pears, chocolate/cocoa, cherries, and pears as shown in Figure 2, however higher concentrations are found in black, green and oolong teas and specifically in green tea due to dried and steamed leaves instead of fermentation as it converts catechin into complex flavonoids(Yang et al., 2018; Shirakami et al., 2016; Khan et al., 2018).A study was conducted by Zhijinget al., 2018 for the characterization of EGC in wine lees in extension as well as terminal units and found that Epigallocatechin was the chief extension unit in white wine lees, however, was not available in the terminal unit of lees.

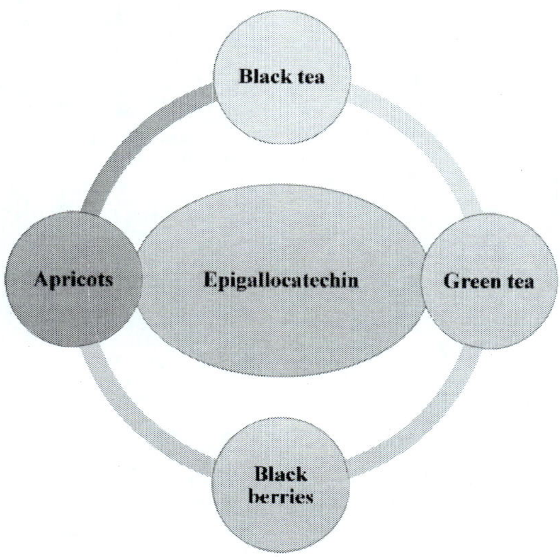

Figure 2. Sources of epigallocatechin.

Arts IC and co-workers, 2000 applied HPLC with fluorescence and UV for the detection was used to determine the levels of (-)-epigallocatechin (EGC), in juices of apple, grapes, iced tea, chocolate milk, coffee, beer, black tea as well red and white wines and concluded that EGC was predominant in tea and was in a negligible amount in wines, juices, chocolate milk, and iced tea. Also, EGC was absent in beer and coffee. The traceable amounts of EGC were found in Artemisia plants but lacked antimalarial properties (Min-Jun Kim et al., 1998).

3. Biological Properties and Therapeutic Potential of Epigallocatechin

Catechins of green tea have numerous health benefits such as free radical scavenging also helps to nullify the effects of pollution and ultraviolet radiation by hampering the degradation the extracellular matrix caused by these two factors (Shi et al., 2016). Catechins also activate collagen synthesis and affect the skin cells (Arct et al., 2003). Epigallocatechin is known to have great therapeutic potential as shown in Figure 3.

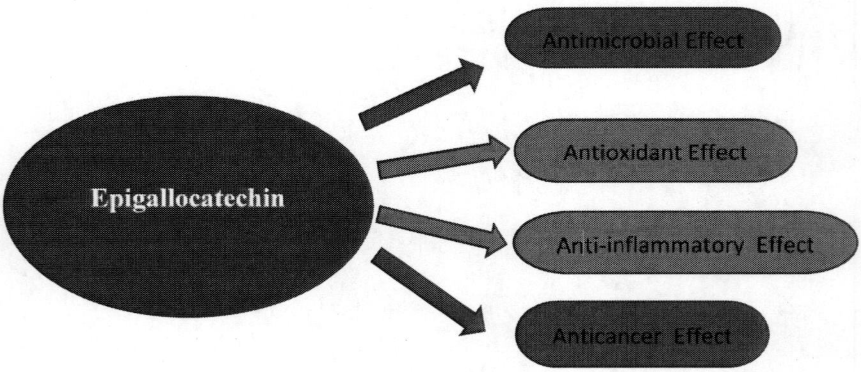

Figure 3. Therapeutic potential of epigallocatechin.

3.1. Antimicrobial Potential of EGC

Bacterial infections which are resistant to the antibiotic are the biggest threat to human beings (Wu et al., 2021). Numerous causalities have been reported to date due to antibiotic-resistant infection, a report by United States Centres for Disease Control and Prevention reported at least 36,000 deaths in the United States in the year 2019 (U.S. Department of Health and Human Services, 2019). Many deaths are reported with methicillin-resistant strains (Cosgrove et al., 2003). Thus, there is an immediate need for alternative antibiotics to reduce the infections. Green tea has an antimicrobial effect on bacteria. Extracts of green tea were reported to reduce the MIC of

Streptococcus mutans (*S. mutans*) in saliva and dental biofilms/plaques of children (Goyal et al., 2017). Catechins are the natural compounds proved to be potent against the infections caused by antibiotic-resistant bacteria (Song et al., 2017). It has been also shown to have effects against a variety of viruses (e.g., HIV, influenza, herpes simplex) (Jigisha et al., 2012; Steinmann et al., 2013). Catechins of green tea have antioxidative action which helps in exerting anti-obesity and anti-inflammatory effects in humans (Bernatova et al., 2018; Terao et al., 1997; Chen et al., 2017).Catechins particularly EGC were recently used by researchers for antimicrobial therapy and the prevention of potent infections. They exhibited different mechanisms against microbes. Catechin EGC has antimicrobial effects against a wide variety of microorganisms (Yam et al., 1997; Taguri et al., 2004, 2006; Taylor et al., 2005; Friedman, 2007; Song and Seong, 2007). EGC is found in the highest amounts in green tea, it is excreted in bile as well in urine, confirming its antimicrobial ability in UTIs (Yang et al., 1998; Lee et al., 2002; Cabrera et al., 2006; Luo et al., 2006).

One study reported (Reygaert and Jusufi, 2013) that the susceptibility of *E.coli* strains (99%) isolated from UTIs was tested and found to be susceptible to $\leq$0.72 mg of EGC levels and suggest growth of *E. coli* in UTIs be controlled up to 6 h with one cup of tea and multiple cups in a day could provide a longer effect. This study concluded that the excretion of EGC in urine is responsible for its antimicrobial effects on *E. coli*. Another study conducted by (Neyestani et al., 2007) for the positive interaction between EGC and antimicrobial agents (gentamicin and amikacin) against *E. coli*.

In another study, the green tea extracts and particles were compared for antibacterial activity. For EGC, a high peak at 269 nm was observed with 20K-P's which were studied for their interaction with *S. mutans*. However, the removal of 20K-P's showed negligible activity and concluded that extracts are better than particles of green tea, and concluded that EGC has the highest antimicrobial resistance (Gopal et al., 2016). One study by (Yam et al., 1997; Hara, 2001) evident the antibacterial activity of the catechin components of green tea and found that EGC is the most potent antibacterial agent. Yam et al., 1998 studied that the catechins have the potential to reverse the methicillin-resistance in sensitive strains and reduce this resistance to a low level so that it could not be possible to inhibit the growth of bacteria.

The interaction of catechins with microbes also seems to played the important role in their antibacterial resistance as they adhere and damage the cell membrane. The relative affinity of catechins for the membranes is reflect their ability to harmonize the β-lactam resistance in Methicillin-resistant *Staphylococcus aureus* (MRSA strains) (Hashimoto et al., 1999, Stapleton et al., 2004a). A study by Stapleton et al., 2006 showed that EGC (nongalloylated) alone cannot reduce the MIC value of oxacillin in mecA-containing strains of *Staphylococcus aureus* but significantly increased the ability of ECg and EGCg to reduce levels of *Staphylococcal* oxacillin resistance. The ECG and EGCG can easily penetrate phosphatidyl-ethanolamine and phosphatidylcholine bilayers, however, EGC is located very close to the phospholipid water interface (Caturla et al., 2003).

Epigallocatechin (EGC) was studied to kill *Escherichia coli* ATCC11775 strain completely in the presence of copper (Hoshino et al., 1999). This bactericidal activity of EGC was complete inhibited by ethylene-diaminetetra acetate, bathocuproine, catalase, superoxide dismutase (SOD), heated catalase, and heated SOD in the presence of copper. Under aerobic conditions, EGC generates hydrogen peroxide with Cu (II) (Hoshino et al., 1999). The generation of hydrogen peroxide could be one reason for its bactericidal action (Arkawa et al., 2004). The hydrogen peroxide can damage the cytoplasmic membrane of *E. coli and* the extent of damage to the membrane was determined by the cellular potassium level (Kohen et al., 1988) as well as cellular ATP level (Aronovitch et al., 1991; Kohen et al., 1988). However, EGC alone was less effective (Kimura et al., 1998).

The bilayer-modifying potency of EGC was tested using gramicidin A (GA) channels and found that EGC increases gramicidin A channel function, with a 500-fold difference in potency (Ingolfsson et al., 2011).

3.2. Antioxidant Potential of EGC

Catechins have phenolic hydroxyl groups to stabilize the free radicals (Fraga et al., 2010). The presence of these groups in catechins is key responsible for their antioxidant activities. Many studies are being conducted to prove the antioxidant potential of tea catechins (Amarowicz & Shahidi, 1995; Wanasundara & Shahidi, 1996; Shahidi & Alexander, 1998; Zhong & Shahidi, 2011; 2012; Zhong, Ma & Shahidi, 2012; Shahidi & Zhong, 2015; McKay & Blumberg, 2002; Perera, Ambigaipalan & Shahidi, 2018; Sun, Zhou & Shahidi, 2018). Antioxidant effects come from the ability of green tea to limit the number of free radicals by binding to reactive oxygen species (ROS) (Serafini et al., 2011; Jigisha et al., 2012; Gupta et al., 2014). Catechins follow different mechanisms for scavenging free radicals such as direct mechanisms by metal ion chelation scavenging reactive oxygen species (ROS) and indirect mechanisms by using antioxidant enzymes and retardation of activity of prooxidant enzymes (Youn et al., 2006). Insufficient antioxidant levels are proved to be connected with diabetes, Alzheimer's disease (Braicu et al., 2013; Shammas et al., 2006). Catechins played a protective and preventive role against oxidative stress (Braicu et al., 2013; Zaveri et al., 2005).

EGC is proved to be a strong antioxidant. The effect of epigallocatechin (EGC) on ARE-mediated gene expression (a pathway for regulation of antioxidants) in the human monocytic cells was investigated (Richard et al.,2008) and found that EGC activated HO-1, GCLM as well as Nrf2 in THP-1 cells, but not NQO1.The silencing of Nrf2 inhibited the expression by EGC, evidencing its importance in the pathway. HO-1 plays a defensive role in vascular diseases (Chung et al., 2008; Yet et al., 2003). One study reported that the oral administration of EGC leads to high antioxidant activity than other gallates (Van Amelsvoort et al., 2001).Another study reported the non-activation of ARE-driven reporters in HepG2 cells by EGC (Chen et al., 2000). The antioxidant ability of EGC was also proved by Kaneko et al., 1993 and reported that four weeks of washing of teeth with catechin solution reduces the halitosis and further supported by Yasuda and Arakawa (1995) who investigated that catechins (EGCG, ECG, and EGC) deodorized methyl mercaptan which indicates the ability of EGC to act as an antioxidant. Few studies (Rice-Evans et al., 1995; Intra and Kuo, 2007; Zaveri et al., 2006; Azman et al., 2014; Fujisawa et al., 2005) reported the radical scavenging efficiency of catechins in the range of relative hierarchy with EGCG at the top position followed by ECG and EGC. EGC promoted lipid oxidation when the efficiency of catechins as an antioxidant was checked in different lipid systems (Huang, Shu-Wen, et al., 1997).

One study was reported (Colon et al., 2012) in which two green tea extracts (GTE1 and GTE2) were prepared with 20% and 75% of total catechins respectively and suggested that EGC is the main catechin that is responsible for the antioxidant capacity in extract in GTE2.In this study, EGC showed a significant change in values after reaction with reactive species as it has more ability to capture free radicals. EGC showed good antioxidant potential than EGCG and ECG when tested against artificial phenothiazine radical cations (Salah et al., 1995). Also, EGC acts as a better antioxidant as compared with epicatechin in bulk corn oil (Huang, Shu-Wen, et al., 1997). Another study reported the antioxidant potential of EGC from tea extract using an online HPLC DPPH assay and concluded that it shows the positive correlation between oxygen radical absorbance capacity and EGC (Chunjian et al., 2014). Catechin metabolites showed different antioxidant activity than their precursors when studied in cultured cells (Spencer et al., 2001) e.g.,

Bacterial ring fusion metabolites of EGC has been reported in human urine and plasma more than their precursors(Li et al., 2000).Organic green tea and conventional tea was compared for amounts of catechins as well as their antioxidant potential and found that organic green tea was having more catechins in them including EGC, ECG, EC, and EGCG which is linked with higher antioxidants in organic green tea (Kismet al., 2015). The antioxidant potential is linked with anti-aging properties. The antiaging ability of EGC was proved by Shin et al., 2018 in which EGC scavenged the methylglyoxal (MGO) which is key precursor information of advanced glycation end products which cause age- and age-related disorders. In another study, by Ambigaipalan et al. (2020), EGC was lyophilized with fatty acids and their derivatives were studied for antioxidant activities in vitro and concluded that enhancement of lipophilicity of EGC with fatty acids leads to increase in its antioxidant potential. The antioxidant potential of these catechins of green tea led to scavenging the free radicals and reactive oxygen species as well as chelating the metal ions, thus reducing the uptake of low-density lipoproteins (LDL) by macrophages as well as preventing its oxidation, hence, in turn, reducing the ability of LDL to aggregate (Satue et al., 1997; Scalbert et al., 2005). EGC in the presence of copper or iron ions was reported to show the antioxidant ability on hydroxyl radicals (Froester et al., 2011).

3.3. Anticancer Potential of EGC

Green tea catechins have many health benefits including anti-cancer (McKay & Blumberg, 2002) which make them very attractive for clinical studies. Animal models have been extensively used for the carcinogenesis studies of skin, lung, stomach, liver as well as esophagus by using tea extracts and inhibition was checked (Yang et al., 1993; Stoner et al., 1995; Katiyar et al., 1996; Yang, 1996). All these catechins are having anti-mutagenic activities, metal chelating as well as radical scavenging activity (Serafini et al., 1996; Wang 1989; Kuroda 1999; Rice et al., 1996) which enlightens their potential to be used against cancer. Many studies have been conducted to prove that EGC and other theaflavins have anti-carcinogenic activity (Lea et al., 1993; Yang 1997;Valcic et al., 1996). It imposed inhibitory effects when studied against human lung tumor cell lines such as H661 and H1299 (Yang et al., 1998). In another study, EGC inhibited the growth of MCF-7 and MDA-MB-231 breast cancer cells when applied with a specific dosage but showed no effect on normal epithelial cells of the breast (Valcic et al., 1996).

In vitro studies were conducted on gender-specific cancer cell lines in humans to examine the effects of EGC on them. The growth of cell lines HH870, DU145, HH450, and HH639 with and without treatment of EGC was studied (Ravindrnath et al., 2006). EGC is recommended as a good candidate against breast cancer as it showed negative effects on cancerous fibroblasts well as transformed keratinocytes, without affecting the normal cells (Chen et al., 1998, Ahmad et al., 1997). EGC was also reported as an inhibitor of xanthine oxidase during the breakdown of purines in hepatic cells (Aucamp et al., 1997). EGC was also studied to inhibit the PC-9 cells (lung cancer cells of humans) (Okabe et al., 1997). EGC was extracted from leaves of green tea and was tested against the HT-29 colon cancer cell line (Valcic et al., 1996). EGC was studied for its anti-proliferative activity against MK-1 cancer cells of the human stomach (Kinjo et al., 2002). Human oral squamous cancer cell line SCC-25 was significantly inhibited by EGC, the inhibition of growth of cells as well as DNA synthesis was dose-dependent (Elattar et al., 2000).Tan et al., 2000 treated LoVo colon cancer cells with EGC which inhibited its growth and started apoptosis by arresting the cells at G1-phase.The antitumor activity of EGC was also proved by checking its inhibitory effects on cancer cells DU145 of the human prostate and inducing apoptosis in them (Chung et al., 2001).

3.4. Preventive Role in Cardiovascular Diseases

Anti-proliferative activity of EGC was reported in which it was shown to inhibit the proliferation of vascular smooth muscle cells and showed a protective role in preventing cardiovascular diseases. It also hampered the activity of platelet-derived growth factor by inhibiting its phosphorylation (Sachinidis et al., 2002). EGC was also studied for its ability to enhance the production of prostacyclin in bovine aortic endothelial cells (Mizugaki et al., 2000). Prostacyclin indirectly inhibits the accumulation of calcium in VSMCs (Huang et al., 1998; Alvarez et al., 2006).This fact was supported by another finding by Furuyashika et al., 2004 which showed the ability of EGC to reduce lipid accumulation intracellularly. Another study reported that the treatment of macrophages with EGC delayed the process of lipid peroxidation and preserving α-tocopherol content of the LDL (Zhu et al., 1999).

3.5. Anti-Inflammation

Inflammation is involved in many conditions such as cardiovascular diseases, arthritis, aging as well as cancer (Serafini et al., 2011; Jigisha et al., 2012). The anti-inflammation potential of green tea was well cited by many researchers (Aneja et al., 2004; Nag et al., 2005; Tipoe et al., 2007; Cavet et al., 2011). The high anti-inflammation effect of green tea is associated with the higher flavonoids in them (Chatterjee et al., 2012) and anti-inflammation is a result of higher production of IL-10 which is an anti-inflammatory cytokine. Many studies revealed that EGC has the potential to be taken as an anti-inflammatory agent. A study by Fechtner et al., 2017 reported that EGC interferes with the IL-1β signaling pathway which regulates IL-6, IL-8, and Cox-2 in rheumatoid arthritis in humans by inhibiting TAK-1 activity. TAK1 plays a key role in inflammatory signaling pathways (Fechtner et al., 2016). EGC was reported to reduce the cytotoxicity of airborne fine dust particles by acting as an anti-inflammatory agent (Kim et al., 2021).

Conclusion and Future Perspectives

Green tea has numerous benefits due to the presence of catechins in it and catechins have broad applications in terms of clinical importance particularly epigallocatechin which was reported to be an antibacterial, anti-inflammatory agent, anti-cancer agent, as well as showed defensive role in cardiovascular diseases. Its antioxidant potential as well as reversing the sensitivity of microbes makes it unique. But despite such benefits, EGC has got little attention of researchers yet to be explored more particularly to be used as an anti-aging agent.

Acknowledgments

Author (SL) is grateful to the University Institute of Biotechnology, Chandigarh University, Gharuan for providing a platform for study and valuable guidance. Authors (MY, NS) are helpful to Head, Maharishi Markandeshwar (Deemed to be University), Mullana-Ambala for necessary help and support.

References

Ambigaipalan, P., Oh, W.Y. and Shahidi, F. (2020). Lipophilized epigallocatechin (EGC) and itsderivatives : Inhibition of oxidation of β-carotene–linoleateoil-in-water emulsion and DNA strand scission. *Journal of Food and Drug Analysis,* 28(3), 356-364.

Arts, I.C., van De Putte, B. and Hollman, P.C. (2000). Catechin contents of foods commonly consumed in The Netherlands. 2. Tea, wine, fruit juices, and chocolate milk. *Journal of Agricultural and Food Chemistry,* 48(5) :1752-1757.

Arct, J., Bielenda, B., Oborska, A. and Pytkowska, K. (2003). The tea and its cosmetic application. *Journal of Applied Cosmetology*, 21 :117–127.

Arakawa, H., Maeda, M., Okubo, S. and Shimamura, T. (2004). Role of hydrogenperoxide in bactericidal action of catechin. *Biological and Pharmaceutical Bulletin,*27, 277–281.

Aronovitch, J. Godinger, D. and Czapski, G. (1991) Bactericidal activity of catecholamine copper complexes. *Free Radical Research Communications*, 12-13 :479–488.

Amarowicz, R., Pegg, R.B., Rahimi-Moghaddam, P., Barl, B. and Weil, J.A. (2004). Free-radical scavenging capacity and antioxidantactivity of selected plant species from the Canadian prairies. *Food Chemistry*, 84(4), 551–562.

Azman, N.A.M., Peiró, S., Fajarí, L., Julià, L. and Almajano, M.P. (2014). Radical scavenging of white tea and its flavonoid constituents by Electron Paramagnetic Resonance (EPR) Spectroscopy. *Journal of Agricultural and Food Chemistry*, 62(25), 5743–5748.

Ahmad, N., Reyes, D.K., Agarwal, R., Mukhtar, H., and Nieminen, A.-L. (1997). Green tea constituent epigallocatechin-3-gallate and induction of apoptosis and cell cycle arrest in human carcinomacells. *JNCI Journal of the National Cancer Institute,* 89(24), 1881–1886.

Aucamp, J.P., Hara, Y. and Apostolides, Z. (2000). Simultaneous analysis of tea catechins, caffeine, gallicacid, theanine, and ascorbicacid by micellar electrokinetic capillary chromatography. *Journal of Chromatography A,* 876(1-2), 235–242.

Alvarez, E., Campos-Toimil, M., Justiniano-Basaran, H., Lugnier, C. and Orallo, F. (2006). Study of the mechanisms involved in the vasorelaxation induced by (−) -epigallocatechin-3-gallate in rat aorta. *British Journal of Pharmacology*, 147(3), 269–280.

Aneja, R., Odoms, K., Denenberg, A.G. and Wong, H.R. (2004). Theaflavin, a black teaextract, is a novel anti-inflammatory compound. *Critical Care Medicine*, 32(10), 2097–2103.

Bungau, S., Abdel-Daim, M.M., Tit, D.M., Ghanem, E., Sato, S., Maruyama-Inoue, M., Yamane, S. and Kadonosono, K. (2019). Health Benefits of Polyphenols and Carotenoids in Age-Related Eye Diseases. *OxidativeMedicine and Cellular Longevity,* 2019, 1-22.

Bradfleld, A.E., Penney, M. and Wright, W.B. (1947). 7. The catechins of green tea. *The Journal of the Acoustical Society of America,* 103, 3−5.

Bradfield, A.E. and Bate-Smith, E.C. (1950). Chromatographic behavior and chemical structure II. The tea catechins. *Biochimica et Biophysica Acta,* 4, 441–444.

Bernatova, I. (2018). Biologicalactivities of (−) -epicatechin and (−) -epicatechin-containing foods: Focus on cardiovascular and neuropsychological health. *Biotechnology b Advances,* 36(3), 666–681.

Braicu, C., Ladomery, M.R., Chedea, V.S., Irimie, A. and Berindan-Neagoe, I. (2013). The relationship between the structure and biological actions of green tea catechins. *Food Chemistry*, 141(3), 3282–3289.

Bhattacharya, S., Chandra, S., Chatterjee, P. and Dey, P. (2012). Evaluation of anti-inflammatory effects of green tea and black tea: A comparative in vitro study. *Journal of Advanced Pharmaceutical Technology & Research*, 3(2), 136-138.

Cabrera, C., Artacho, R. and Giménez, R. (2006). Beneficia lEffects of Green Tea—A Review. *Journal of the American College of Nutrition*, 25(2), 79–99.

Casas, R., Castro-Barquero, S., Estruch, R. and Sacanella, E. (2018). Nutrition and Cardiovascular Health. *International Journal of Molecular Sciences*, 19(12),3988.

Cosgrove, S.E., Sakoulas, G., Perencevich, E.N., Schwaber, M.J., Karchmer, A.W. and Carmeli, Y. (2003). Comparison of mortality associated with methicillin resistant and methicillin susceptible *Staphylococcus aureus* A Meta-analysis. *Clinical Infectious Diseases*, 36(1)53–59.

Chen, J.-H., Lee, M.-S., Wang, C.-P., Hsu, C.-C. and Lin, H.-H. (2016). Autophagic effects of *Hibiscus sabdariffa* leaf polyphenols and epicatechin gallate (ECG) against oxidized LDL-inducedinjury of human endothelial cells. *European Journal of Nutrition,* 56(5), 1963–1981.

Cabrera, C., Artacho, R. and Giménez, R. (2006). BeneficialEffects of Green Tea—A Review. *Journal of the American College of Nutrition*, 25(2), 79–99.

Chung, S.W., Liu, X., Macias, A.A., Baron, R.M. and Perrella, M.A. (2008). Heme oxygenase-1–derived carbon monoxide enhances the host defense response to microbial sepsis in mice. *Journal of Clinical Investigation,* 118(1), 239–247.

Chen, C., Yu, R., Owuor, E.D. and Tony Kong, A.-N. (2000). Activation of antioxidant-response element (ARE), mitogen-activated protein kinases (MAPKs), and caspases by major green tea polyphenol components during cell survival and death. *Archives of Pharmacal Research*, 23(6), 605–612.

Colon, M. and Nerin, C. (2012). Role of Catechins in the Antioxidant Capacity of an Active Film Containing Green Tea, Green Coffee, and Grapefruit Extracts. *Journal of Agricultural and Food Chemistry.*

Chen, Z.P., Schell, J.B., Ho, C.-T. and Chen, K.Y. (1998). Green tea epigallocatechin gallate shows a pronounced growth inhibitory effect on cancerous cells but not on their normal counterparts. *Cancer Letters*, 129(2), 173–179.

Cavet, M.E., Harrington, K.L., Vollmer, T.R., Ward, K.W. and Zhang, J.Z. (2011) Anti-inflammatory and anti-oxidative effects of the green tea polyphenol epigallocatechin gallate in human cornealepithelial cells. *Molecular Vision.* 2011; 17 :533–42.

Chung, L.Y., Cheung, T.C., Kong, S.K., Fung, K.P., Choy, Y.M., Chan, Z.Y. and Kwok, T.T. (2001). Induction of apoptosis by green tea catechins in human prostate cancer DU145 cells. *Life Sciences*, 68(10), 1207–1214.

Elattar, T.M. and Virji, A.S. (2000). Effect of tea polyphenols on growth of oral squamous carcinoma cells *in vitro. Anticancer Research,* 2000 ; 20 :3459–3465.

Fraga, C.G., Croft, K.D., Kennedy, D.O. and Tomas-Barberan, F.A. (2019). The effects of polyphenols and other bioactives on human health. *Food and Function,* 10(2), 514–528.

Fung, S.-T., Ho, C.K., Choi, S.-W., Chung, W.-Y. and Benzie, I.F.F. (2012). Comparison of catechin profiles in human plasma and urine after single dosing and regular intake of green tea (*Camellia sinensis). British Journal of Nutrition,* 109(12),2199–2207.

Friedman, M. (2007). Overview of antibacterial, antitoxin, antiviral, and antifungal activities of tea flavonoids and teas. *Molecular Nutrition & Food Research*, 51(1), 116–134.

Fraga, C.G., Galleano, M., Verstraeten, S.V. and Oteiza, P.I. (2010). Basic biochemical mechanisms behind the health benefits of polyphenols. *Molecular Aspects of Medicine*, 31(6), 435–445.

Fujisawa, S. and Kadoma, Y. (2006). Comparative study of the alkyl and peroxy radical scavenging activities of polyphenols. *Chemosphere*, 62(1), 71–79.

Forester, S.C. and Lambert, J.D. (2011). The role of antioxidant versus pro-oxidant effects of green tea polyphenols in cancer prevention. *Molecular Nutrition & Food Research*,55(6), 844–854.

Fechtner, S., Singh, A., Chourasia, M. and Ahmed, S. (2017). Molecular insights into the differences in anti-inflammatory activities of green tea catechins on IL-1β signaling in rheumatoid arthritis synovial fibroblasts. *Toxicology and applied pharmacology*, 329, 112–120.

Fechtner, S., Fox, D.A., Ahmed, S. (2017). Transforming growth factor β activated kinase 1: a potential therapeutic target for rheumatic diseases. *Rheumatology* (Oxford), 1;56(7) :1060-1068.

Furuyashiki, T.,Nagayasu, H., Aoki, Y., Bessho, H., Hashimoto, T., Kanazawa, K. And Ashida, H. (2004). Tea catechin suppresses adipocyte differentiation accompanied by down-regulation of pparγ2 and c/ebpα in 3t3-11 cells. *Bioscience, Biotechnology, and Biochemistry*, 68(11), 2353–2359.

Gupta, D.A., Bhaskar, D.J., Gupta, R.K., Karim, B., Jain, A., and Dalai, D.R. (2014). Green tea: are view on its natural anti-oxidant therapy and cariostatic benefits. *Issues in Biological Sciences and Pharmaceutical Research*.2, 8–12.

Goyal, A., Bhat, M., Sharma, M., Garg, M., Khairwa, A. and Garg, R. (2017). Effect of green tea mouthrinse on Streptococcus mutans in plaque and saliva in children : An in vivo study. *Journal of Indian Society of Pedodontics and Preventive Dentistry*, 35(1), 41-46.

Gopal, J., Muthu, M., Paul, D., Kim, D.-H. and Chun, S. (2016). Bactericidal activity of green tea extracts : the importance of catechin containing nano particles. Scientific Reports, 6(1), 19710.

Hara, Y. (2001). Green tea: health benefits and applications. New York, USA : Marcel Dekker.

Hashimoto, T., Kumazawa, S., Nanjo, F., Hara, Y. and Nakayama, T. (1999). Interaction of tea catechins with lipid bilayers investigated with liposome systems. *Bioscience, Biotechnology, and Biochemistry*, 63(12), 2252–2255.

Huang, S.-W. and Frankel, E.N. (1997). Antioxidant activity of tea catechins in different lipid systems. *Journal of Agricultural and Food Chemistry,* 45(8), 3033–3038.

Huang, Y., A. Zhang, C.W. Lau, and Z.Y. Chen. (1998). Vasorelaxant effects of purified green tea epigallocatechin derivatives in rat mesentericartery. *Life Sciences.*63 **:**275-283.

Ingólfsson, H.I., Koeppe, R.E. and Andersen, O.S. (2011). Effects of green tea catechins on gramicidin channel function and inferred changes in bilayer properties. *FEBS Letters,* 585(19), 3101–3105.

Intra, J. and Kuo, S.-M. (2007). Physiological levels of tea catechins increase cellular lipid antioxidant activity of vitamin C and vitamin E in human intestinal Caco-2 cells. *Chemico-Biological Interactions,* 169(2), 91–99.

Jin, Y., Jin, C.H. and Ho Row, K. (2006). Separation of catechin compounds from different teas. *Biotechnology Journal,* 1(2), 209–213.

Jigisha, A., Nishant, R., Navin, K., and Pankaj, G. (2012). Green tea :a magical herb with miraculous outcomes. *International Research Journal of Pharmacy,* 3, 139–148.

Khan, N. and Mukhtar, H. (2018). Tea polyphenols in promotion of human health. *Nutrients,* 11(1), 39.

Kohen, R. and Chevion, M. (1988). Cytoplasmic membrane is the target organelle for transition metal mediated damage induced by paraquat in *Escherichia coli. Biochemistry,* 27(7), 2597–2603.

Kimura, T., Hoshino, N., Yamaji, A., Hayakawa, F. and Ando, T. (1998). Bactericidal activity of catechin-copper (II) complexes on *Escherichia coli* ATCC11775 in the absence of hydrogenperoxide. *Letters in Applied Microbiology,* 27(6), 328–330.

Kaneko, K., Shimano, N., Suzuki, Y., Nakamukai, M., Ikazaki, R., Ishida, N., Kanayasu, E., Karuda, T., Takihara, T., Sakane, I., Yayabe, F., and Matsui, T. (1993). Effects of tea catechins on oral odor and dental plaque. *Oral Therapeutics and Pharmacology,* 12 :189–197.

Kim, M.J., Kim, J.H., Kim, J.H. and Kim, Y.J. (2015). Comparative studies on the antioxidant capacities and catechin profiles of conventional and organic green tea. *Journal of the Korean Society for Applied Biological Chemistry*, 58(4), 475–480.

Katiyar, S. and Mukhtar, H. (1996). Tea in chemoprevention of cancer. *International Journal of Oncology*,8, 221–238.

Kuroda, Y. and Hara, Y. (1999). Antimutagenic and anticarcinogenic activity of tea polyphenols. *Mutation Research/Reviews in Mutation Research*, 436(1), 69–97.

Kinjo, J., Nagao, T., Tanaka, T., Nonaka, G., Okawa, M., Nohara, T. and Okabe, H. (2002). Activity-guided fractionation of green tea extract with antiproliferative activity against human stomach cancer cells. *Biological and Pharmaceutical Bulletin*, 25(9), 1238–1240.

Kim, J., Choi, H., Choi, D.-H., Park, K., Kim, H.-J. and Park, M. (2021). Application of green tea catechins, polysaccharides, and flavonol prevent fine dustinduced bronchial damage by modulating inflammation and airway cilia. *Scientific Reports,* 11(1) 2232.

Kim, M.J. and Lee, C-H. (1998). The effects of extracts from Mugwort on the blood ethanollevel and liver function. *Korean Journal for Food Science of Animal Resources*, 18(4), 348-357.

Lee M.J., Maliakal P., Chen L., Meng X., Bondoc F.Y., Prabhu S., et al. (2002). Pharmacokinetics of tea catechins after ingestion of green tea and (-) -epigallocatechin-3-gallate by humans : formation of different metabolites and individual variability. *Cancer Epidemiology, Biomarkers & Prevention*. 11, 1025–1032.

Luo, H., Tang, L., Tang, M., Billam, M., Huang, T., Yu, J., Wei, Z., Liang, Y., Wang, K., Zhang, Z.-Q., Zhang, L. and Wang, J.-S. (2005). Phase IIachemoprevention trial of green tea polyphenols in high-risk individuals of liver cancer : modulation of urinary excretion of green tea polyphenols and 8-hydroxydeoxyguanosine. *Carcinogenesis*, 27(2), 262–268.

Li, C., Lee, M.-J., Sheng, S., Meng, X., Prabhu, S., Winnik, B., Huang, B., Chung, J.Y., Yan, S., Ho, C.-T. and Yang, C.S. (2000). Structural Identification of two metabolites of catechins and their kinetics in human urine and blood after tea ingestion. *Chemical Research in Toxicology*, 13(3), 177–184.

Lea, M.A., Xiao, Q., Sadhukhan, A.K., Cottle, S., Wang, Z.-Y. and Yang, C.S. (1993). Inhibitoryeffects of teaextracts and (−) -epigallocatechin gallate on DNA synthesis and proliferation of hepatoma and erythroleukemia cells. *Cancer Letters*, 68(2-3), 231–236.

Matute, A., Tabart, J., Cheramy-Bien, J.-P., Pirotte, B., Kevers, C., Auger, C., Schini-Kerth, V., Dommes, J., Defraigne, J.-O. and Pincemail, J. (2020). Compared phenolic compound contents of 22 commercial fruit and vegetable juices : relationship to ex-vivo vascular reactivity and potential In vivo projection. *Antioxidants*, 9(2), 92.

McKay, D.L. and Blumberg, J.B. (2002). The role of tea in human health : An update. *Journal of the American College of Nutrition*, 21(1), 1–13.

Mizugaki, M., Ishizawa, F., Yamazaki, T. and Hishinuma, T. (2000). Epigallocatechin gallate increase the prostacyclin production of bovine aortic endothelial cells. *Prostaglandins & Other Lipid Mediators*, 62(2), 157–164.

Neyestani, T.R., Khalaji, N. and Gharavi, A. (2007). Selective microbiologic effects of tea extract on certain antibiotics against *Escherichia coli* in vitro. *The Journal of Alternative and Complementary Medicine*, 13(10), 1119–1124.

Nagchaudhuri, A., Karmakar, S., Roy, D., Pal, S., Pal, M. and Sen, T. (2005). Anti-inflammatory activity of Indian black tea (Sikkim variety). *Pharmacological Research*, 51(2), 169–175.

Ogborne, R.M., Rushworth, S.A. and O'Connell, M.A. (2008). Epigallocatechin activates haem oxygenase-1 expression via protein kinase Cδ and Nrf2. *Biochemical and Biophysical Research Communications*, 373(4), 584–588.

Okabe, S., Suganuma, M., Hayashi, M., Sueoka, E., Komori, A. and Fujiki, H. (1997). Mechanisms of growth inhibition of human lung cancer cell line, PC-9, by tea polyphenols. *Japanese Journal of Cancer Research*, 88(7), 639–643.

Potì, F., Santi, D., Spaggiari, G., Zimetti, F. and Zanotti, I. (2019). Polyphenol health effects on cardiovascular and neurodegenerative disorders : A review and meta-analysis. *International Journal of Molecular Sciences*, 20(2), 351.

Perera, N., Ambigaipalan, P. and Shahidi, F. (2018). Epigallocatechin gallate (EGCG) esters with different chain lengths fatty acids and their antioxidant activity in food and biological systems. *Journal of Food Bioactives*, 1(1), 124-133.

Rodríguez-García, C., Sánchez-Quesada, C. and Gaforio, J.J. (2019). Dietary Flavonoids as Cancer Chemopreventive Agents : An Updated Review of Human Studies. *Antioxidants*, 8(5), 137.

Reygaert, W.C. (2014). The antimicrobial possibilities of green tea. *Frontiers in Microbiology,* 5 :434.

Reygaert, W. and Jusufi, I. (2013). Green tea as an effective antimicrobial for urinary tract infections caused by *Escherichia coli. Frontiers in Microbiology,* 4, 162.

Rice-Evans, C.A., Miller, N.J., Bolwell, P.G., Bramley, P.M. and Pridham, J.B. (1995). The relative antioxidant activities of plant-derived polyphenolic flavonoids. *Free Radical Research,* 22(4), 375–383.

Rice-Evans, C.A., Miller, N.J. and Paganga, G. (1996). Structure-antioxidant activity relationships of flavonoids and phenolicacids. *Free Radical Biology and Medicine,* 20(7), 933–956.

Ravindranath, M.H., Saravanan, T.S., Monteclaro, C.C., Presser, N., Ye, X., Selvan, S.R. and Brosman, S. (2006). Epicatechins purified from green tea (*Camellia sinensis*) differentially suppress growth of gender-dependent human cancer celllines. *Evidence-Based Complementary and Alternative Medicine*, 3(2), 237–247.

Shirakami, Y., Sakai, H., Kochi, T., Seishima, M. and Shimizu, M. (2016) Catechins and itsrole in chronic diseases. *Advances in Experimental Medicine and Biology,* 929 :67-90.

Shi, M., Nie, Y., Zheng, X.-Q., Lu, J.-L., Liang, Y.-R. and Ye, J.-H. (2016). Ultraviolet B (UVB) photo sensitivities of tea catechins and the relevant chemical conversions. *Molecules,* 21(10), 1345.

Song, M., Teng, Z., Li, M., Niu, X., Wang, J. and Deng, X. (2017). Epigallocatechin gallate inhibits *Streptococcus pneumoniae* virulence by simultaneously targeting Pneumolysin and Sortase A. *Journal of Cellular and Molecular Medicine,* 21(10), 2586–2598.

Steinmann, J., Buer, J., Pietschmann, T. and Steinmann, E. (2013). Anti-infective properties of epigallocatechin-3-gallate (EGCG), a component of green tea. *British Journal of Pharmacology, 1*68(5), 1059–1073.

Song, J.M. and Seong, B.L. (2007). Tea catechins as a potential alternative anti-infectious agent. *Expert Review of Anti-infective Therapy*, 5(3), 497–506.

Stapleton, P.D., Shah, S., Anderson, J.C., Hara, Y., Hamilton-Miller, J.M.T. and Taylor, P.W. (2004). Modulation of β-lactamresistance in *Staphylococcus aureus* by catechins and gallates. *International Journal of Antimicrobial Agents*, 23(5), 462–467.

Stapleton, P.D., Shah, S., Hara, Y. and Taylor, P.W. (2006). Potentiation of catechin gallate-mediatedsensitization of *Staphylococcus aureus* to oxacillin by nongalloylated catechins. *Antimicrobial Agents and Chemotherapy,* 50(2), 752–755.

Shahidi, F. and Alexander, D.M. (1998). Green tea catechins as inhibitors of oxidation of meatlipids. *J. Food Lipid.* 5 : 125-133.

Shahidi, F. and Alexander, D.M. (1998). Green tea catechins as inhibitors of oxidation of meatlipids. *Journal Of Food Lipids*, 5(2), 125–133.

Shahidi, F. and Zhong, Y. (2015). U.S. Patent No. 9,018,248. Washington, DC : U.S. Patent and Trademark Office.

Sun, Y., Zhou, D. and Shahidi, F. (2018). Antioxidant properties of tyrosol and hydroxytyrosol saturated fatty acid esters. *Food Chemistry,* 245, 1262–1268.

Serafini, M., Rio, D., N'Dri, Y., Bettuzzi, S. and Peluso, I. (2011). Health benefits of tea. *Oxidative Stress and Disease*, Chapter 12, 239–261.

Shammas, M.A., Neri, P., Koley, H., Batchu, R.B., Bertheau, R.C., Munshi, V., Prabhala, R., Fulciniti, M., Tai, Y. tzu, Treon, S.P., Goyal, R.K., Anderson, K.C. and Munshi, N.C. (2006). Specific killing of multiple myelomacells by (-) -epigallocatechin-3-gallate extracted from green tea :biologic activity and therapeutic implications. *Blood,* 108(8), 2804–2810.

Salah, N., Miller, N.J., Paganga, G., Tijburg, L., Bolwell, G.P. and Rice-Evans, C. (1995). Polyphenolic flavanols as scavengers of aqueous phase radicals and as chain-breaking antioxidants. *Archives of Biochemistry and Biophysics*, 322(2), 339–346.

Spencer, J.P.E., Schroeter, H., Crossthwaithe, A.J., Kuhnle, G., Williams, R.J. and Rice-Evans, C. (2001). Contrasting influences of glucuronidation and O -methylation of epicatechin on hydrogenperoxide-induced cell death in neurons and fibroblasts. *Free Radical Biology and Medicine*, 31(9), 1139–1146.

Shin, E.R., Jung, W., Kim, M.K. and Chong, Y. (2018). Identification of (−) -epigallocatechin (EGC) as a methylglyoxal (MGO)-trapping agent and thereby as an inhibitor of advanced glycation end product (AGE) formation. *Applied Biological Chemistry*, 61(5), 587–591.

Satué-Gracia, M.T., Heinonen, M. and Frankel, E.N. (1997). Anthocyanins as antioxidants on human low-densitylipoprotein and lecithin−liposome systems. *Journal of Agricultural and Food Chemistry*, 45(9), 3362–3367.

Scalbert, A., Manach, C., Morand, C., Rémésy, C. and Jiménez, L. (2005). Dietary polyphenols and the prevention of diseases. *Critical Reviews in Food Science and Nutrition*, 45(4), 287–306.

Stoner, G.D. and Mukhtar, H. (1995). Polyphenols as cancer chemopreventive agents. *Journal of Cellular Biochemistry,* 59(S22), 169–180.

Serafini M, Ghiselli A, Ferro-Luzzi A. (1996) In vivo antioxidant effect of green and black tea in man. *European Journal of Clinical Nutrition*, 50 : 28–32.

Sachinidis, A., Skach, R.A., Seul, C., Ko, Y., Hescheler, J.R., Ahn, H-Y. and Fingerle, J.R. (2002). Inhibition of the PDGF β-receptortyrosine phosphorylation and its downstream intracellular signal transduction pathway in rat and human vascular smooth muscle cells by different catechins. *FASEB J.*16(8) :893-895.

Terao, J., Piskula, M. and Yao, Q. (1994). Protective Effect of Epicatechin, Epicatechin Gallate, and Quercetin on Lipid Peroxidation in Phospholipid Bilayers. *Archives of Biochemistry and Biophysics*, 308(1), 278–284.

Taylor, P.W., Hamilton-Miller, J.M.T. and Stapleton, P.D. (2005). Antimicrobial properties of green tea catechins. *Food Science Technology Bulletin :Functional Foods*, 2(7),71–81.

Tipoe, G., Leung, T.-M., Hung, M.-W. and Fung, M.-L. (2007). Green teapolyphenols as an anti-oxidant and anti-inflammatory agent for cardiovascular protection. *Cardiovascular & Hematological Disorders-Drug Targets,* 7(2), 135–144.

Tan, X., Hu, D., Li, S., Han, Y., Zhang, Y. and Zhou, D. (2000). Differences of four catechins in cell cycle arrest and induction of apoptosis in LoVocells. *Cancer Letters,* 158(1), 1–6.

Vergote, D., Cren-Olivé, C., Chopin, V., Toillon, R.-A., Rolando, C., Hondermarck, H. and Bourhis, X.L. (2002). (−) -Epigallocatechin (EGC) of green tea induces apoptosis of humanbreast cancer cells but not of their normal counterparts. *Breast Cancer Research and Treatment*, 76(3), 195–201.

Van Amelsvoort, J.M.M., Van Het Hof, K.H., Mathot, J.N.J.J., Mulder, T.P.J., Wiersma, A. and Tijburg, L.B.M. (2001). Plasma concentrations of individual tea catechins after a single oral dose in humans. *Xenobiotica*, 31(12), 891–901.

Valcic, S., Timmermann, B.N., Alberts, D.S., Wachter, G.A., Krutzsch, M., Wymer, J. and Guillen, J.M. (1996). Inhibitory effect of six green tea catechins and caffeine on the growth of four selected human tumor cell lines. *Anti-CancerDrugs,* 7(4), 461–468.

Wu, M. and Brown, A.C. (2021). Applications of catechins in the treatment of bacterial infections. *Pathogens*, 10(5), 546.

Wanasundara, U.N. and Shahidi, F. (1996). Stabilization of seal blubber and menhaden oilswith green tea catechins. *Journal of the American Oil Chemists' Society*, 73(9), 1183–1190.

Wang, Z.Y., Cheng, S.J., Zhou, Z.C., Athar, M., Khan, W.A., Bickers, D.R. and Mukhtar, H. (1989). Antimutagenic activity of green teapolyphenols. *Mutation Research/Genetic Toxicology,* 223(3), 273–285.

Yang, G. (1998). Inhibition of growth and induction of apoptosis in human cancer cell lines by tea polyphenols. *Carcinogenesis,* 19(4), 611–616.

Yang, C.S., Wang, H. and Sheridan, Z.P. (2018). Studies on prevention of obesity, metabolic syndrome, diabetes, cardiovascular diseases and cancer by tea. *Journal of Food and Drug Analysis*, 26(1), 1–13.

Yam, T.S., Shah, S. and Hamilton-Miller, J.M.T. (2006). Microbiological activity of whole and fractionated crude extracts of tea (*Camellia sinensis*), and of tea components. *FEMS Microbiology Letters*, 152(1), 169–174.

Yang C.S., Chen L., Lee M.J., Balentine D., Kuo M.C., Schantz S.P. (1998). Blood and urine levels of tea catechins after ingestion of different amounts of green tea by human volunteers. *Cancer Epidemiology, Biomarkers & Prevention,*7(4):351-354.

Yam, T.S., Shah, S. and Hamilton-Miller, J.M.T. (2006). Microbiological activity of whole and fractionated crude extracts of tea (*Camellia sinensis),* and of tea components. *FEMS Microbiology Letters,* 152(1), 169–174.

Yam, T. (1998). The effect of a component of tea (*Camellia sinensis*) on methicillin resistance, PBP2' synthesis, and beta-lactamase production in *Staphylococcus aureus. Journal of Antimicrobial Chemotherapy*, 42(2), 211–216.

Youn, H.S., Lee, J.Y., Saitoh, S.I., Miyake, K., Kang, K.W., Choi, Y.J. and Hwang, D.H. (2006). Suppression of MyD88- and TRIF-dependent signaling pathways of toll-like receptor by (−) -epigallocatechin-3-gallate, a polyphenol component of green tea. *Biochemical Pharmacology,* 72(7), 850–859.

Yet, S.-F., Layne, M.D., Liu, X., Chen, Y.-H., Ith, B., Sibinga, N.E.S. and Perrella, M.A. (2003). Absence of heme oxygenase-1 exacerbates atherosclerotic lesion formation and vascular remodeling. *The FASEB Journal,* 17(12), 1759–1761.

Yasuda, H., & Arakawa, T. (1995). Deodorizing mechanism of (–) -Epigallocatechin gallate against methyl mercaptan. *Bioscience, Biotechnology, and Biochemistry, 59,* 1232-1236.

Yang, C.S. and Wang, Z.-Y. (1993). Tea and Cancer. *JNCI Journal of the National Cancer Institute,* 85(13), 1038–1049.

Yang, C.S., Chen, L., Lee, M.-J. and Landau, J.M. (1996) Effects of tea on carcinogenesis in animal models and humans (Editedunder the auspices of the American Institute for Cancer Research). In *dietaryphytochemicals in cancer prevention and treatment.* Plenum Press, New York, 51–61.

Yang, G.-Y., Liu, Z., Seril, D.N., Liao, J., Ding, W., Kim, S., Bondoc, F. and Yang, C.S. (1997) Black tea constituents, the aflavins, inhibit 4- (methylnitrosamino)-1-(3-pyridyl) -1 butanone (NNK)-induced lung tumorigenesis in A/J mice. *Carcinogenesis*, 18, 2361–2365.

Yang, G. (1998). Inhibition of growth and induction of apoptosis in human cancer cell lines by tea polyphenols. *Carcinogenesis,* 19(4), 611–616.

Younes, M., Aggett, P., Aguilar, F., Crebelli, R., Dusemund, B., Filipič, M., Frutos, M.J., Galtier, P., Gott, D., Gundert-Remy, U., Lambré, C., Leblanc, J., Lillegaard, I.T., Moldeus, P., Mortensen, A., Oskarsson, A., Stankovic, I., Waalkens-Berendsen, I., Woutersen, R.A. and Andrade, R.J. (2018). Scientific opinion on the safety of green tea catechins. *EFSA Journal,* 16(4) : 5239-5289.

Zhao, C., Li, C., Liu, S., and Yang, L. (2014). The Galloylcatechins to main antioxidant capacity of tea made from *Camellia sinensis* in China. *The Scientific World Journal,* 2014, 1–11.

Zhijing, Y., Shavandi, A., Harrison, R. and Bekhit, A.E.-D. (2018). Characterization of Phenolic Compounds in Wine Lees. *Antioxidants*, 7(4), 48.

Zhong, Y. and Shahidi, F. (2011). Lipophilized Epigallocatechin gallate (EGCG) derivatives as novel antioxidants. *Journal of Agricultural and Food Chemistry,* 59(12), 6526–6533.

Zhong, Y. and Shahidi, F. (2012). Lipophilis edepigallocatechin gallate (EGCG) derivatives and their antioxidant potential in food and biological systems. *Food Chemistry,* 131(1), 22–30.

Zhong, Y., Ma, C.-M. and Shahidi, F. (2012). Antioxidant and antiviral activities of lipophilic epigallocatechin gallate (EGCG) derivatives. *Journal of Functional Foods*, 4(1), 87–93.

Zaveri, N.T. (2006). Green tea and its polyphenolic catechins :Medicinal uses in cancer and noncancer applications. *Life Sciences*, 78(18), 2073–2080.

Zhu, Q.Y., Huang, Y., Tsang, D. and Chen, Z.-Y. (1999). Regeneration of α-Tocopherol in human low-densitylipoprotein by green tea catechin. *Journal of Agricultural and Food Chemistry*, 47(5), 2020–2025.

Chapter 7

Therapeutic Potential of Genistein: Structure, Dietary Source, Current Status and Future Perspectives

Siddhartha Dan[1], Deeksha Sharma[2], Kiran Upadhyay[3] and Sushil Kumar Upadhyay[4,*]

[1]Department of Biotechnology, I.K. Gujral Punjab Technical University, Jalandhar, Punjab, India
[2]Shoolini University of Biotechnology and Management Sciences, Shoolini University, Solan, Himachal Pradesh, India
[3]Department of Zoology, K.P. Higher Education Institute, Jhalwa-Prayagraj, Uttar Pradesh, India
[4]Department of Biotechnology, Maharishi Markandeshwar (Deemed to be University), Mullana-Ambala, Haryana, India

Abstract

In the field of developing new therapeutic agents; the compounds derived from plants recently attracted scientists with greater concentration. The plant-derived compounds are screened widely due to their diverse pharmacological potentials. Phytoestrogens such as soy-derived isoflavones (a well known polyphenols) have been broadly studied because they comprising carcinogenesis inhibition ability. Plant derived compounds are quite similar to the chemical structure of 17β-estradiol and initiate estrogens receptor binding in target organs, to exert estrogenic effects. An isoflavone i.e., Genistein derived (60% of total Isoflavone) from soy rich products. It has been reported that Genistein exhibit various biological effects, such as improvement of glucose

* Corresponding Author's Email: upadhyay.k.sushil@gmail.com.

In: Flavonoids
Editors: Nirmala Sehrawat, Diwakar Aggarwal, Sushil Kumar Upadhyay et al.
ISBN: 978-1-68507-953-6

metabolism, anti-tumour activity (inhibition of cell proliferation, regulation of the cell cycle, induction of apoptosis), reduction of pre-menopausal and postmenopausal hot flashes, impairment of angiogenesis in both hormone-related and hormone-unrelated cancer cells and modulation of antioxidant effects. Furthermore, clinical and epidemiological studies reported number of health benefits of genistein in several chronic diseases, such as diabetes, osteoporosis and cardiovascular disease, and helpful in the enhancement of distinctive menopausal symptoms, including depression and anxiety. Even though the biological effects are more promising having certain restrictions, such as effects on target organs, low bioavailability and biological estrogenic activity, so that it may limited the clinical bids of genistein to some extent. In some studies, it was documented that the modifications in molecular structure of genistein may eliminate the activity of estrogen in the target organs. In the present chapter, authors summarize the structure, dietary source, latent benefits of genistein on consequences on pre-menopause and post-menopause, osteoporosis, anxiety, obesity, cardiovascular, diabetes, breast cancer and depression.

Keywords: genistein, therapeutic potential, isoflavone, anticancer, menopause pharmacokinetics

1. Introduction

In Asian countries, Genistein [5,7-dihyroxy-3-(-4-hydroxyphenyl)-4H-1-benzopyran-4-one] i.e., a major isoflavone - present in soy and souy-based food products are regularly consumed by people (Ronis, 2016). Certainly, the consumption of isoflavones by adults in China and Japan is around 25-50mg which is far greater than the consumption by women in Western countries i.e., 3mg (Sak, 2017a). Genistein is a phytoestrogen compound derived from plants as a main source, and used as traditional food (Milligan et al., 1999). Phytoestrogens are secondary metabolites derived from plants which mimic 17β–estradiol (mammalian estrogen) functionally and structurally both (Rafi et al., 2000; Sak, 2017b; Jaiswal et al., 2018). Consequently, genistein can be used to overwhelm breast cancer as well. Good numbers of phytoestrogens subtypes are known which include lignans, coumestans, isoflavones, flavones, prenylflavonoids and chalcones. The most common phytoestrogen form is the isoflavones (Kim et al., 2014; Ganai and Farooqi, 2015; Spagnuolo et al., 2015; Uifalean et al., 2015). Isoflavones are made of diphenolic structure which found to be identical to the structure of synthetic estrogenshexestrol and diethylstilbestrol. There are two basic isoflavones subgroups i.e., glycosides and aglycones. The Genistein is included in subgroup aglycone of isoflavones (Ravindranath et al., 2004).

Numerous epidemiological studies suggested that, there are lesser incidences of various types of cancer (i.e., prostate cancer and breast cancer) in Asian countries as compared to Western countries (Tuli et al., 2021). Consequently, the recent interest has been focused on the suppression and prevention of tumorigenesis by the possible contribution of isoflavones through high dietary consumption (Farina et al., 2006). In the previous times, numbers of studies have been available about genistein for its potential anticancer role both in animal models and cell cultures (Barnes et al., 2000; Dixon and Ferreira, 2002). The natural isoflavone can explicit a wide range of medicinal properties which includes antiangiogenic, antioxidant, antiproliferative activities, proapoptotic and anti-inflammatory. Above all these aforementioned properties genistein comprises chemotherapeutic and chemopreventive potential (Dixon and Ferreira, 2002; Ososki and Kennelly, 2003; Magee and Rowland, 2004).

Genistein (4'5,7-trihydroxyisoflavone) is placed in family Leguminosae which includes soybean (*Glycine max*) and occurs as a glycoside (Genistin). Among Europeans, Americans and Asians there is a significant correlation between the incidence of gender based cancers and the serum/plasma level of genistein, which suggested that genistein might reduce the risk of tumours formation. Few evidences showed the action mechanism of genistein in cancer cells and normal cells. Genistein inhibits the activity of topoisomerase, which helps in DNA replication, repair and transcription and also inhibits protein tyrosine kinase (PTK) included in tyrosyl residues phosphorylation of receptors bounded to membrane and leads to signal transduction. The genistein can modify the expression to facilitate the immune recognition of gangliosides and carbohydrate antigens as well. Synergistically, genistein acts with various drugs like cisplastin, dexamethasone, tamoxifen, 1,3-bis-2-chloroethyl-1-nitrosourea (BCNU), tiazofurin and daunorubicin and with some bioflavonoid food supplements which includes black-tea Thearubigins, green-tea Catechins and Quercetin (Matsuda et al., 2001).

2. Structure of Genistein

Structurally, 5,7-dihydroxy-3-(4-hydroxyphenyl) chromen-4-one (Genistein) or 4,5,7-Trihydroxyisoflavone is included in family isoflavone and obtained from soy products like soybeans. Genistein is a widely known secondary metabolite from plants consists of 3-phenylchromen-4-one nucleus made of two aromatic rings (A and B). Additionally, these two rings are linked to another pyran ring (C). There is another functional group present in its carbon skeleton are double bonded C_2-C_3 and at C_4 position an oxo group at ring C. Furthermore, three hydroxyl groups are present at positions C_5, C_7 and C_4' of A and B rings respectively. The Genistein was first isolated from a flowering plant *Genista tinctoria* (Dyer's broom) of family Fabaceae (Perkin and Newbury, 1899). The Genistein was also successfully documented in 1928 for the first time from many plants like fava beans, kudzu, soyabeans and lupin (Baker and Robinson, 1928; Sehrawat et al., 2020; Yadav et al., 2020; Yadav et al., 2021a,b). Genistein derivatives also be achieved by Ferrier rearrangements of 3,4-di-O-acetyl-L-rhamnal with 3-bromopropanol to 2,3-unsaturated bromoalkylglycosides after epoxide with the mCPBA (meta-chloroperoxybenzoic acid) and the subsequently associated with the genistein (Gojet al., 2012). Moreover, derivatives of genistein could also be derived through novel glycoconjugation and chemical glycosylation process (Szeja et al., 2017). The main dietary source of biologically active glucosidegenistein, digestion or fermentation of soy products and soya beans, which results in the release of sugar molecules from glycoside isoflavones and leaving aglyconeisoflavone, genistein (Markovits et al., 1989). Genistein comprises diphenol structure which stereo-chemically resembles to human endogenous estrogens (E2) (Tham et al., 1998; Cassidy, 1999). On opposite sides of genistein, the similar distance between OH group and E2 molecules makes genistein capable of binding between subtypes α and β of ER (estrogen receptor).

The molecular formula of Genistein is $C_{15}H_{10}O_5$ and the storage temp is 20°C, however, the melting point about 297–298°C. The molecular weight of Genistein is 270.24, found to be DMSO soluble, and water insoluble with light sensitive off-white powder form. Genistein is [4'5,7-trihydroxyisoflavone or 5,7-dihydroxy-3-(4-hydroxyphenyl)chromen-4-one] ($C_{15}H_{10}O_5$) which is included in natural multifunctional isoflavonoid class of flavonoids having C_{15} skeleton (Yang et al., 2012). Genistein is an example of phytoestrogenic compound which is similar to other plant compounds like lignans having estrogenic effect. First time genistein was isolated in 1899 from *Genista tinctoria* L. and named after this plant genus. After that, it is found that genistein occurs in *Glycine max* L. (*Soja hispida*) and *Trifolium* species as a secondary metabolite (Polkowski and Mazurek, 2000). Chemically, the structure of genistein and estradiol is similar and leads the binding to ERs

(Kuiper et al., 1998; Klinge, 2000; Kurzer, 2002; Yoon et al., 2014). In polar solvents, Genistein have high solubility such as in DMSO (Dimethylsulfoxide), ethanol and acetone, but as compared to water the solubility is less. Even though, in 1928 the organic synthesis of genistein also obtained using different methods. Genistein chemical synthesis was performed by cyclization of corresponding ketones using microwave oven method (Chang et al., 1994). In elicitor-treated soybean cell suspension cultures, which are, biotechnological synthesis was achieved by converting (2S)-naringin into genisteinin NAD(P)H- oxygen-dependent conditions and by the addition of cytochrome P-450 (CYP) (Kochs and Grisebach, 1986). Using *Saccharomyces cerevisiae* (yeast cells) which contain isoflavone synthase gene i.e., isolated from *Glycyrrhyza echinara* L., a metabolic engineering approach was set up for the synthesis of genistein (Katsuyama et al., 2007). Accordingly, genistein was produced in *Nicotiana tabacum* L. leaves transformed with isoflavone synthase, by acting at the phenylpropanoid metabolism, although UV-B treatment also increased the genistein production in *Arabidopsis* species (family Brassicaceae) (Yu et al., 2000). Because of several beneficial effects of genistein, the pharmacists are encouraged to synthesize genistein derivatives having improved pharmacologic profile. There are numerous genistein derivatives are reported right now which includes 7-O-heterocycle (Meng et al., 1999), 7-O-b-D-glucuronic acid (Somjen et al., 2002), halogenated (Matsumoto et al., 2005; Zengin, 2011-), hydroxylated (Zhang et al., 2007), 7-O-b-D-glucoside (Kgomotso et al., 2008), FA-esterified (Fuet al., 2008), nitroxy (Shimoda et al., 2008), glycoconjugate and alkylbenzylamine (Choi et al., 2009), 6-carboxymethyl (Zheng et al., 2010), dimethylaminomethyl (Fan et al., 2011), benzylated (Hyz et al., 2011), benzosulfonate (Li et al., 2012), phenoxyalkyl carboxylic acid (Qiang et al., 2014) and esterified (Maniewska et al., 2019), and derivatives. It was revealed from various data that main features of genistein structure overlay the synthesis profiling of new derivatives of genistein which can be emerged as innovative and profound types of anti-osteoporetic, anticancer and estrogenic agents.

3. Dietary Source of Genistein

Soybean (without cholesterol) is accounted as a high protein containing vegetable and contains a predominant ingredient, the Genistein. Very small amount of genistein is found in various legumes, for e.g., chickpeas (garbanzo beans). Genistein is present in variable amount in edible soy based food materials, for e.g., tofu, soy flour, soy milk, tempeh, soy protein isolates, miso and soy based newborn child formulas. About 53% of soy proteins are present in the soy flour. A meat replacement i.e., Textured soy protein (TSP) is prepared from defatted soy which is present in hot dogs, meat loafs, sausages, hamburgers, contain 50-70% of soy protein, which usually depends upon initially utilized soy material. Soy protein isolates (SPI) contain 90% of soy protein which is used in formation of special nutritional foods (e.g., energy bars, sports drink, and special diets) for unwell and body building beverages. The other plant based foods that contain genistein are barley meal, sunflower, alfalfa, broccoli, clover seeds, clover sprouts, caraway and cauliflower. The range of genistein concentration is 0.2 to 1.0mg/g. It was found that genistein can also be present in *Trifolium* species and isolated from different kind of microorganisms (*Streptomyces spp.* and *Pseudomonas spp.*) fermentation broth (Ganai and Farooqi, 2015).

Soy- based foods are best known sources for genistein, like soy drinks (i.e., soy-based beverages or soy milk) or soy cheese. The amount of genistein present in mature soybeans is 5.6 to 276.0mg/100g and average content is 81.0mg/100g that have been described for various purposes (Bhagwat et al., 2008). Soy foods also contain other isoflavone which is different from genistein by the absence of one hydroxyl group at 5th position i.e., known as daidzein. Both of the isoflavones exist in glycoside or aglycone form. The most common forms are presence of O-β-D-glucoside derivatives at 7th positions in both daidzein and genistein. Traditionally, there are number of Asian foods made from soybeans, so the average dietary intake of isoflavone lies between the range of 25.0-50.0mg/day, but in western countries, the estimated average intake of isoflavone was less than 2.0mg/day (VanErp-Baart et al., 2003; Messina et al., 2006). The second most important source of genistein is legume with a range between 0.2 to 0.6mg/100g, and present together with daidzein (Liggins et al., 2000a,b). A typical example of legume is the genus *Lupinus* which widely cultivated for seeds because it has same nutritional value as the soybean. Other than soybeans, few important legumes include chick peas and broad beans, which contain significant amount of genistein as well. The amount of genistein varies considerably in nuts, vegetables and fruits and the predictable range of genistein documented between 0.03 to 0.2mg/100g (Liggins et al., 2000). Biotechnological approach is the best way to maximize the yield of isoflavonoid through which the medicinal and nutritional value of various foods can be improved. The metabolic processes of seed germination, the degradation of food reserves and the characteristic anabolic processes dedicated to the sprouting of embryo, have been shown to increase the nutritional value primarily by enhancing the content of plant secondary metabolites i.e., isoflavonoids and vitamins (Ahmad and Pathak, 2000). In soybeans and soy-related products, the increased amount of genistein and other aglycones isoflavones are well documented (Quinhone and Ida, 2015). During fermentation process of soybean products, the amount of genistein and other aglycones were found to be increased (Lee et al., 2014). It is possible through genetic manipulations. The genistein can also be obtained from non-leguminous plant sources for example; rice. In transgenic rice, isoflavone synthase (IFS) was cloned from genistein-rice soybean cultivator that contain 30-fold more increased amount of genistein (Sohn et al., 2014). In United States and Europe, soy milk, soy yogurt, soy-based meat substitutes and soy cheese, have gained popularity due to medicinal significance of genistein and other isoflavonoids.

The most popular wellsprings of genistein are soy-based food sources including soy cheddar or soy drinks (i.e., soy milk and soy-based refreshments). However, in some local cherry cultivars of Hungarian beginning, genistein recorded up to 4.4 mg/100g. An all-inclusive rundown of food sources with their genistein content is accessible online in a few data sets (Bhagwat et al., 2008). The biotechnological approach used to boost the isoflavonoid yield by growing seeds is the commonest technique used to work on the healthful and restorative upsides of specific food sources. The metabolic cycles of seed germination, have been displayed to upgrade healthy benefit essentially by expanding the substance of nutrients and plant optional metabolites, for example, isoflavonoids (Ahmad and Pathak, 2000; Kim et al., 2005; Bhagwat et al., 2008; Yuan et al., 2009; Paucar-Menacho et al., 2010; Shi et al., 2010). As needs be, the expanded substance of genistein and other isoflavonoid aglycones has been very much recorded in developed soybean seeds and related items (Quinhone and Ida, 2015.

4. Bioavailability of Genistein

The mean volume of distribution normalised to bioavailability (Vd/F), clearance rate, and half-life of genistein in the human investigation using radio labelled isoflavones were 258.76 L, 21.85 L/h, and 7.77 h, respectively (Yang et al., 2012; Chandrasekharan and Aglin, 2013). Isoflavone bioavailability refers to the amount of these chemicals that become available for tissue distribution and physiologic effects. Genistein from soy extracts, its free form, and its glycoside genistin have been found to be highly accessible in a variety of experimental scenarios including *in vivo* experiments. Genistein was rapidly accessible and detected in portal vein plasma just after 15 minutes of treatment in freely moving unanesthetized rats using a cannula in the portal vein, with AUC values (0–24 h) of 54 and 24mmol h/kg, respectively (Steensma et al., 2006; Mazumder and Hongsprabhas, 2016). Several investigations have documented that genistin showed a better oral bioavailability than genistein (Kwon et al., 2007). The weak water solubility of genistein resulted its bioavailability to be limited after oral ingestion (Motlekar et al., 2006; de Oliveira, 2016). Genistein also has a bitter taste to necessitating formulations (Table 1) that address both the bioavailability and the acceptable flavour issues (Huang et al., 1982; 1992). Both people and experimental animals have also shown that genistein undergoes extensive metabolism in the gut and after absorption. Dihydrogenistein, dihydrodaidzein, 69-hydroxy-Odesmethylangolensin, 4-ethylphenol, glucuronoide and sulphate conjugates of genistein and its metabolites, and 4-hydroxyphenyl-2-propionic aid are one of the metabolite found in the blood and excreta (de Oliveira, 2016). The C-ring of the isoflavonoid skeleton is recognized to be cleaved by the gut microbiota, giving 4-hydroxyphenyl-2-propionic acid and dihydrogenistein (Tamura et al., 2007).

Table 1. Genistein formulation through potential preparation methods for multiple applications

Type and main components of formulations	Methods of preparation and applications	Reference(s)
Tablets (Sodium croscarmellose, Crospovidone, Hydroxypropyl methylcellulose, Sodium dodecyl sulfate)	• Wet granulation and direct compression method • Immediate release of genistein and diadzein	Horter and Dressman, 2001
Solid lipid microparticles(Glycerylpalmitostearate, L-α-lecithin, Tween 80)	• Melt dispersion technique • Enhancement of bioavailability	Desai and Thakkar, 2016
Microparticles(Gelatin A)	• Spray drying technique • Enhanced aqueous solubility and bioavailability	Zhang et al., 2003
Solid lipid Nanoparticles (Glycerylpalmitostearate, L- α - lecithin, Poloxamer 188)	• Hot homogenization method • Enhancement of bioavailability	Desai and Thakkar, 2016
Hydrogel Matrix (*k*-Carrageenan)	• Matrix retention method • Improve the water solubility of genistein	Kim et al., 2017

Type and main components of formulations	Methods of preparation and applications	Reference(s)
Micelles (Pluronic F127)	• Solid dispersion method • Improve oral bioavailability of genistein	de Oliveira et al., 2013
Nanoparticles (Eudragit E, Poloxamer 188) and (Poly (Ɛ-caprolactone). D-α tocopheryl polyethylene glycol 1000 succinate)	• Nanoprecipitation method • Improve water solubility and bioavailability of genistein	Tosi et al., 2007
	• Modified nanoprecipitation method • Enhanced permeability and retention effects of genistein NPs in tumor cells	Mohanraj et al., 2006
Nanostructured lipid carriers (Glycerol monostearate, Soy lecithin)	• High-speed homogenization and ultrasonication probe method • Enhanced oral bioavailability of poorly bioaccessible molecules	Barratt, 2003
Layered Nano emulsions (Chia seed oil, Chitosan, Tween 80)	• Homogenization and ultrasonication method • Targeted delivery along with extended release of genistein	Tang et al., 2011
Topical Nano emulsions (Medium chain triglycerides and Octyldodecanol)	• Spontaneous emulsification method • Permeation enhancement of genistein through skin.	Zhang et al., 2015

5. Pharmacokinetics of Genistein

Genistein is metabolized from Biochanin A i.e., a plant precursor parent compound. When isoflavones are present as glycosides in bound form, then they found to be inactive, but the compound becomes active after the removal of sugar residues. Genistein with the help of intestinal microflora undergoes fermentation; then both aglycone (unfermented parent) and metabolites are reliable for absorption (Sakai and Kogiso, 2008). The administration of genistein, orally, results with t_{max} of 5-6h and $t_{1/2}$ of 8h for the absorption of genistein (Setchell et al., 2001; Setchell et al., 2003; Chandrasekharan and Aglin, 2013). During oral administration of genistein, only 20-40% is absorbed through gastrointestinal tract, in which it undergoes for enterohepatic cycling. Thus, oral absorption of genistein is very low through GI tract, which leads to low bioavailability (Zheng et al., 2010). Genistein is distributed rapidly to all tissues and crosses the blood brain barrier and placental barriers as well. The highest distribution of genistein is found in liver tissues and gastro-intestinal tract, which consistent with enterohepatic recycling.

6. Remedial Competence of Genistein

6.1. Metabolic Pathway and Activity

Genistein underwent as an inhibitor of the tyrosine-explicit protein kinases of the epidermal development factor (EGF) receptor and furthermore hindered the movement of topoisomerases (Kim et al., 1998). Genistein conceivably hinders multiplication of different disease cells and prompts cell separation and apoptosis. Genistein actuated G2/M capture has been discovered to be related with upregulated p21 articulation in prostate, lung and breast cancer (Chan et al., 2018). Genistein (10–100μM; 10min) additionally diminished glucose up-take in both estrogen receptor-positive MCF-7 and - negative (MDA-MB-231) breast malignancy cell lines have been summarized in Table. 2 (Keating and Martel, 2018). In the meantime, genistein additionally showed antioxidant and antiangiogenic activities that were significant for malignant growth avoidance. Perhaps the main benefits of genistein are its low toxicity in correlation with numerous current chemotherapeutic medications. The isoflavones are likewise utilized in avoidance of heart sicknesses (Food Standards Agency 2003; Kurzer, 2003). The preventive cardiovascular impacts of genistein are unclear, yet might be because of its cell reinforcement movement, bringing down of serum cholesterol, restraint of tyrosine kinase and additionally improvement of vascular reactivity (Fotsis et al., 1993; Potter, 1995). Genistein have critical bone saving impacts among postmenopausal ladies (Potter, 1995). In vitro and in vivo contemplates proposed the valuable impacts on bone mineral thickness and bone turnover. In any case, the specific instruments are baffling and temporary, yet an isoflavone rich eating regimen seems to offer a high potential for the anticipation of osteoporosis (Setchell and Lydeking, 2003; Setchell et al., 2003).

6.2. Antimetastasis Effect

The cancer metastasis is perhaps the leading reason for death around the world. The metastasis of malignancy is discovered to be reliant upon higher articulation of MMPs. In an examination utilizing naked mice, genistein was found to hinder the metastasis of salivary adenoid cystic carcinoma cells (ACC) by restraining the VEGF and MMP-9 articulation (Liu and Yu, 2004). Further confirmations recommended that genistein showed an inhibitory impact on the movement of MAT-LyLu and AT-2 rodent prostate disease cells (Miekus and Madeja, 2007).

6.3. Anti-Inflammatory and Antioxidant Effect

It is understood very well that development of cancer is closely related to inflammatory behaviour. Variety of inflammatory mediators, such as chemokines and cytokines are expressed by cancer cells. For the treatment of inflammatory responses, anticancer drugs can be used, is suggested by literature. It is reported in previous studies that genistein posed inhibitory effect on signal transducer and activator of transcription 1 and in activated macrophages, NF-kB medicated signalling cascade (Hamalainen et al., 2007). *In vivo* and *in vitro* studies of mammary glands of female rats, it was reported that the inflammatory responses found suppressed using genistein by inhibiting protein kinase R-like ER kinase, Cox-2, pp38 and phosphorylated c-Jun N-terminal kinase (JNK) (Hwang et al., 2009). Down regulation of inflammatory mediators like TNF-α, NF-kB, IL-1β, IL-8 and IL-6 were also reported in studies done using MH7A human synoviocyte cells and RAW 264.7 macrophages (Ji et al., 2012; Li et al., 2015). Moreover, through inactivation of toll like receptor-4 and NF-kB, the expression of cytokines involved in inflammatory responses like TNF-α and IL-1β, were suppressed by genistein (Jeong et al., 2014). In murine model of peritoneal endometriosis, the administration of genistein of dose 1.04 or 1.3mg/day, reduces the expression of IL-6 and TNF-α, which indicated that genistein can inhibit the inflammation development (Tuli et al., 2019). Recently, researchers reported that the mice in which hepatocellular carcinoma induced by diethylnitrosamine, after long term treatment of genistein showed anticancer and anti-inflammatory responses (Tuli et al., 2019).

7. Genistein Metabolites and Cancer

Despite the fact, genistein is accounted for to be utilized fundamentally through oxidation, sulfation, glucuronidation, hydroxylation, or methylation the impact of genistein metabolites on its anticancer property does not seen obviously (Nguyen et al., 2006). Metabolites, for example, 5,7,39,49-tetrahydroxyisoflavone (THIF) and 2 glutathinyl forms of THIF were recognized in T47D tumorigenic bosom epithelial cells treated with genistein. Since, the THIF has been displayed to hinder angiogenesis and endothelial cell expansion (Kiriakidis et al., 2005). It is advantageous to take note of that the arrangement of THIF during genistein treatment may assume a significant part in cell cycle capture, hindrance of cell multiplication, and actuation of flagging pathways, for example, p38 MAPK, which seen in T47D cells. Besides, oxidation of THIF to o-quinone alongside development of hydrogen peroxides and receptive oxygen species instigates DNA strand breakage. This prompts the initiation of the ataxia telangiectasia and Rad3-related kinase (ATR) flagging pathway, which enacts the kinases associated with DNA harm registration control (Vauzour et al., 2007).

7.1. Epidemiology of Liver Cancer

Recently, in Japan a case-control study investigated a population-based prospective, in which the role of estrogens in the development of liver cancer, which is then verified that whether isoflavones are associated with risks of liver cancer (Michikawa et al., 2015). Patients are selected by authors having infection of hepatitis B or C virus at baseline and the concentration of isoflavones (daidzein, equol, glycitein and genistein) in plasma. It was indicated in the study in middle aged men and women of Japan having hepatitis virus infection did not showed association between the occurrences of risk of liver cancer (Spagnuolo et al., 2015).

7.2. Epidemiology of Gastric Cancer

From interventional view point, the role of soybean against gastric cancer always remains a matter of debate. In Korean multicenter cancer cohort, a case-control study suggested that decreased risk of gastric cancer were associated with high concentration of isoflavones (Ko et al., 2010). On the contrary, in Japan Public Health Center-Based prospective, a parallel case study indicated that among Japanese women and men, there is null association between risk of gastric cancer and intake of isoflavone (Hara et al., 2013).

7.3. Epidemiology of Lung Cancer

In tumorigenesis, it has been shown that estrogens have mitogenic effects in lung cancer cells and they interact with pathways with growth factor; conflicting results are produced in epidemiologic studies reflected association between intake of isoflavone and risks of lung cancer (Shimazu et al., 2010).

Though, in Asia prospective studies are carried out, that indicated inverse association in non-smokers (Spagnuolo et al., 2015). A nested case–control study within a high prevalence of non-smokers linked plasma isoflavone concentration, a large-scale population-based prospective study with different isoflavone intakes in Japanese women, and lung cancer risk demonstrated an inverse relationship as well (Shimazu et al., 2011).

7.4. Epidemiology of Colorectal Cancer

In animal and human studies, it is found that consumption of soy reduces risk of colon cancer. Evidences from epidemiological studies indicated that phytoestrogens protect against colorectal cancer development (Rossi et al., 2006; Spagnuolo et al., 2015). For example, among healthy patients and people belongs to population- based Ontario Familial Colorectal Cancer Registry, there is a relation between risks of colorectal cancer and intake of dietary phytoestrogen (lignans, total phytoestrogens and isoflavones). There was an association between intake of dietary isoflavone and lignin with the reduction of risk of colorectal cancer significantly. Furthermore, polymorphic genes encoding for enzymes involved in phytoestrogens metabolism [catechol O-methyl transferase, UDP- glucuronosyl transferase (UGT), CYP, glutathione S-transferases (GSTs)] were not subjected for modification with respect to intake of phytoestrogens during observation (Cotterchio et al., 2006).

7.5. Epidemiology of Breast Cancer

An inverse relationship was showed between risk of breast cancer and intake of soy in various case-control studies (Spagnuolo et al., 2015). In Japan, a prospective study cohort found that consumption of isoflavone and miso soup frequently was associated with reduction of breast cancer risk (Yamamoto et al., 2003). Evidently, the isoflavone content of soybeans and soy-containing products is related to chemo-preventive effects of these products (Spagnuolo et al., 2015)

Table 2. The molecular targets modulated to achieve anti-cancer activity of genistein

Cancer-Cell line (Genistein Concentration)		Molecular targets modulated to achieve anti-cancer activity	Reference(s)
BREAST	Cell line MCF-7 (50 μM)	• Nuclear factor kappa B (NFκB), AKT, BRCA1, BRCA2, human epidermal growth factor receptor 2, epidermal growth factor receptor, platelet-derived growth factor receptor, LRP, Abl. • Decrease HER2 expression, apoptosis increases suppressor proteins.	Hsieh et al., 1998; Chen, et al., 2003; Gu et al., 2005
	Cell line MDA-MB-231 (30μM)	• Nuclear factor kappa B, AKT, cyclin-dependent kinase inhibitor 1/CIP1, G1 Phase. • Decrease phosphorylation of AKT and increases nuclear factor kappa B DNA-binding activity, MDM-2-mediated degradation of phosphoprotein p53, and p21WAFI	Zhou et al., 2001a,b; Satoh et al., 2003
LIVER	Cell line HepG2 (10–20μM)	• Transforming growth factor beta, NFAT1, focal adhesion kinase, epidermal growth factor receptor, G2/M phase, nuclear factor kappa B, mitogen-activated protein kinase, phosphatidylinositol-3-kinase and protein kinase B. • Cell cycle arrest, decrease migration, MAPK, PI3K/AKT signalling pathways and apoptosis	Chodon et al., 2007
	Cell line Bel-7402 (10 μg/mL)	• p125FAK, G0/G1 and G2/M phase. • Increases cell cycle arrest in the G0/G1 and G2/M phase, Decrease p125FAK	Chodon et al., 2007; Li et al., 2017
	Cell line HuH-7 (20 μM)	• Caspase -3, -6, -7, -8, -10, matrix metalloproteinase 9, nuclear factor kappa B, mitogen-activated protein kinase and phosphatidylinositol-3-kinase and protein kinase B. • Increases apoptosis, fragmentation of DNA, Decrease nuclear factor kappa B activity	Mansoor et al., 2011; Wang et al., 2014

Cancer-Cell line (Genistein Concentration)		Molecular targets modulated to achieve anti-cancer activity	Reference(s)
	Cell line Hep3B (15–25μM)	• p38 mitogen-activated protein kinase, cysteine aspartic acid specific protease, nuclear factor kappa B, phosphoprotein p53, mitogen-activated protein kinase. • AMPK-mediated anti-inflammation and pro-apoptosis, decrease TNF and IL-6, apoptosis, fragmentation of DNA, increases endoplasmic reticulum stress and mitochondrial insult	Yeh et al., 2007; Mansoor et al., 2011; Wang et al., 2014; Li et al., 2017; Lee et al., 2019
	Cell line SMMC-7721 (10–20μM)	• Cysteine aspartic acid specific protease, nuclear factor kappa B, G2/M phase, transforming growth factor beta, mitogen-activated protein kinase, phosphatidylinositol-3-kinase and protein kinase B, phosphoprotein p53. • Apoptosis, decrease of nuclear factor kappa B activity, increases cell cycle arrest in the G2/M phase.	El-Serag et al., 2008; Hong et al., 2008;Tian et al., 2014
LUNG	Cell line A549 (25–50μM)	• Epidermal growth factor receptor, nuclear factor kappa B, G2/M, miR-27a, MET. • Cell cycle arrest, apoptosis, G2 phase arrest, decrease MET protein expression levels, increases apoptosis via miR-27a and MET signalling pathways.	Mukherjee et al., 2010; Shiau et al., 2010; Yang et al., 2016
	Cell line H446 (25 μM)	• Fork-head box protein M1, Cdc25B, cyclin B, surviving. • Decrease FoXM1, Cdc25B, cyclin B and survivin, increases apoptosis.	Tian et al., 2014
GASTRIC	Cell line BGC-823 (25 μM)	• BCL2-associated X protein, B-cell lymphoma 2, NF-κB, Cox-2, G2/M, caspase-3, AKT. • Apoptosis, Decrease Bcl-2, cell proliferation, G2/M Phase arrest, breakdown of caspases.	Yang et al., 2016; Thangavel et al., 2019
	Cell line SGC-7901 (10–20 μg/mL)	• Extracellular signal-regulated kinases 1 and 2 (MAPK1/3) phosphatidylinositol-3-kinase and protein kinase B, protein tyrosine phosphatase, Ser642, small cell protein, a mitotic inhibitor kinase, Cdc2/Cdk1, Thr15. • Decrease tyrosine-specific protein kinases, phosphorylation of EP300 by inhibiting the activity of MAPK1, increases apoptosis.	Thangavel et al., 2019

Table 2. (Continued)

Cancer-Cell line (Genistein Concentration)		Molecular targets modulated to achieve anti-cancer activity	Reference(s)
CORECTAL	DLD-1 Cell line (75 μM)	• Nuclear β-catenin, phospho-β-catenin, secreted frizzled related protein 2, Wingless/integrated pathway. • Decrease β-catenin-mediated Wingless/integrated signalling through increasing sFRP2 gene.	Thangavel et al., 2019
	Cell Line SW480 (10 μM)	• p21, cyclin D1, C- myelocytoma, dickkopf Wingless/integrated signalling pathway inhibitor 1. • Increases mRNA and Dickkopf Wingless/integrated signalling pathway inhibitor 1protein levels, decrease cell proliferation, Induce histone acetylation.	Arai et al., 2000; Thangavel et al., 2019
	Cell line HT29 (60–120 μM)	• G2/M and S phases, cyclin-dependent kinase inhibitor 1, B-cell lymphoma 2-associated X protein / B-cell lymphoma 2, phase, forkhead box O3. • G2/M phase cell cycle arrests, increases apoptosis through B-cell lymphoma 2 family proteins, cyclin-dependent kinase inhibitor 1during the cell cycle.	Zhu et al., 2002; Yu et al., 2004; Pool et al., 2018
	Cell line HCT116 (25–50 μmol/L)	• Metalloproteinase, vascular endothelial growth factor 3, forkhead box O3, phosphoprotein p53, phosphatidylinositol-3-kinase and protein kinase B, G2/M phase. • Silencing of phosphoprotein phosphor protein p53-determined activity of fork head box O3, induce G2/M phase cell cycle arrest and apoptosis, Decrease MMP-2 and Fms-related tyrosine Kinase 4.	Qi et al., 2011; Xiao et al., 2015
	Cell line SW1116 (10–30 μg/mL)	• G2/M Phase, protein tyrosine kinases, topoisomerase-II, proteoglycans/glycosaminoglycans. • Cell cycle arrest in G2/M phase, decrease protein tyrosine kinases and topoisomerase II, affects the synthesis of proteoglycans / glycosaminoglycans and decrease cell proliferation.	Chatzinikolaou et al., 2007; Yan et al., 2010

8. Clinical Trials

In human clinical trials, some controversial results confirmed, in which, there were some cases when genistein consumed at 54mg/d dose, it did not showed any adverse estrogenic effect on breast tissue, whereas supplementation of dietary soy on breast tissue reflected proestrogenic effects (Squadrito et al., 2013). The use of genistein with ER-positive breast cancers in women, after considering agonist activity against ER-a of genistein, must be considered carefully. In this regard, there are 2 clinical trials based on the use of genistein in breast cancer, a phase I analysis titled "Genistein in Preventing Breast or Endometrial Cancer in Healthy Postmenopausal Women," has been completed, although results have not yet been published and a phase II investigate titled "Gemcitabine Hydrochloride and Genistein in Treating Women with Stage IV Breast Cancer" (Spagnuoloet al., 2015).

9. Therapeutic Potential of Genistein

Genistein shows twenty times greater selectivity for estrogen receptors (ER) β than α and is estrogenic isoflavone, which can be a beneficial property of phytoestrogens. It considered that most of the side effects are in association with estrogens which established by the binding with ER-α, while the beneficiary effects found to associate with ER-β binding. Genistein has inhibitory activity which promises to a latent candidate used for hormone replacement therapy (HRT), on various symptoms of post-menopause as well as breast and ovarian cancer. Genistein emulate different effects of estrogens by acting on various molecular pathways, without knowing the life-threatening side effects (Sakla et al., 2007; Pintova et al., 2019). Still, the probability of genistein for the treatment of symptoms of post-menopause is unexplored. The activation of biological pathways by genistein is reported in several studies and its mode of action on animal models and cell lines and its use in HRT as well (Sakla et al., 2007). The action of genistein is produced by targeting different enzymes like, ERs, protein tyrosine kinases (PTK) ERs, mitogen-activated protein kinase (MAPK) A, topoisomerase I and II ERs, ATP-binding cassette (ABC) transporters, 5α-reductase, protein histidine kinase, and peroxisome proliferator-activated receptor-gamma (PPAR-γ) ERs, amid others (Huang et al., 1992; Evans et al., 1995; Du et al., 2018; Thangavel et al., 2019). Genistein also shows some other health benefits which includes the reduction of postmenopausal symptoms (vaginal dryness, hot flashes), reducing the incidences of cardiovascular disease and prevention of osteoporosis (Marini et al., 2007; Si and Liu, 2007). Genistein also plays role in obesity by reducing appetite, and then decreasing fat tissue deposition and body mass. Genistein ingestion alters the function and concentration of different hormones like thyroid hormones, by reducing thyroid peroxidase activity (Chang and Doerge, 2000), insulin, by overwhelming the effect of insulin-stimulation of oxidation of glucose and antilipolytic effect (Nogowski et al., 2002) and leptin, by showing its effect on adipose tissue receptors and by decreasing its secretion. In adrenal cortical cells, genistein is shown to decrease cortisol production and by the inhibition of cytochrome P450-21hydroxylase and adrenocortical 3β-hydroxysteroid dehydrogenase (Ohno et al., 2002; 2003). The various therapeutic effects of genistein are summarized in Table 3.

Table 3. Therapeutic effects of genistein on some diseases

Disease	Genistein Effects	Reference(s)
Hyperarousal	Improves serotonin metabolism, stabilizes monoamine oxidase activity, and improves 5-hydroxy indole acetic acid/serotonin turnover ratio	Thangavel et al., 2019
Vasomotor	Decreased frequency of hot flashes, night sweats, and sleep disturbances; as well as symptoms of depression and memory loss	Mareti et al., 2019; Thangavel et al., 2019
Diabetes mellitus	Decreased fasting glucose concentration, insulin resistance, and improved Glycemic index	Guo et al., 2019; Thangavel et al., 2019
Circulatory system	Reduction of the cell death of cardiomyocytes, serum levels of tumour necrosis factor-alpha and macrophages, the severity of atherosclerosis, and Acute myocardial infarction incidence	Mareti et al., 2019; Thangavel et al., 2019
Carcinoma	Decreased the case incidence of lung, ovarian, breast, hepatocellular and gastric carcinoma	Thangavel et al., 2019
Bariatric patients	Decreased serum concentration of total cholesterol, low-density lipoprotein, triglycerides and high-density lipoprotein	Thangavel et al., 2019

9.1. Effect of Genistein in Treating Postmenopausal Vasomotor Symptoms

The essential symptoms of estrogen deficiency during menopause are insomnia, hot flashes and night sweats and vasomotor symptoms (Gartoulla et al., 2016; Caretto et al., 2018). Though, most of the women experiences these symptoms during post-menopause and perimenopause results in affecting the quality of life, emotional unleashing and some other problems (such as anxiety, fatigue, depression, lack of focus and mood swings) (Thangavel et al., 2019). In Latin America and Europe, about 70-80% of women experiences hot flashes during post-menopause whereas in Asia, only 20% women have same experience. In Japan, a community-based study was conducted by Messina et al. (2006) having women with intake of isoflavone, consumption of soy products and hot flashes incidences (Thangavel et al., 2019). The study showed that intake of soy and its products reduce the hot flash incidences as compared to the beginning of study, over a span of six years. It is concluded in a study that isoflavone and soy-derived product consumption reduces the hot flash incidences, even in low amount (75.2-115.9g/day). It was also reported that women with high soy intake have better response rates than women with low soy intake (Thangavel et al., 2019). In a clinical research, it was found that genistein administration of dosage 30mg for 12 weeks, resulted reduction of hot flashes by 51% (9.4-4.7/day), in a double-blind study, whereas in placebo group the reduction by 27% (9.9-7.1/day) (Thangavel et al., 2019). Although the method by which genistein lowers hot flashes has not been fully explained and it has been speculated that genistein is a good ER modulator, it may act in cells by the traditional genetic mechanism, entering cells by diffusing through the lipid bilayer. The nucleus, mRNA synthesis, and tissue-specific protein creation are all stimulated by this complex event (Thangavel et al., 2019). Findings from various studies suggested that genistein can be used to upgrade vasomotor and emotional symptoms. Additional studies can be done where it will be found out that whether genistein can be used to find promising ways for the treatment of menopause associated symptoms.

9.2. Cardioprotective Effects of Genistein

Due to estrogen loss, cardiovascular disease incidences are increased substantially in postmenopausal women (Thangavel et al., 2019). In postmenopausal women, the risks of HRT were demonstrated in the studies done by World Women's Health initiative. According to these studies HRT should not have been continued in postmenopausal women, (Teede, 2007), because it is linked to an increased risk of breast, ovarian, and endometrial malignancies, uterine haemorrhage, venous thromboembolism, and stroke.

Thus, alternate approaches, like consumption of isoflavones from fruits, plants and soybeans, may provide health benefits less or no side effects among women showing post-menopausal symptoms (Sehrawat et al., 2020; Yadav et al., 2020; Yadav et al., 2021a,b). These products contain genistein in high amount, which shown to reduce menopausal symptoms, improve heart function and increases bone mineral density (Atteritano et al., 2007; Si and Liu, 2007). Some other studies suggested that genistein exert multiple actions in insulin secretion on islet cells of pancreas and being an inhibitor of protein tyrosine kinase (Irace et al., 2013). In Langerhans islet cells, genistein increases basal secretion of insulin and without affecting metabolism of glucose, it inhibits sulfonylurea-stimulated insulin release. A study was conducted by Crisafulli et al. (2005) on administering 54mg/day dosage of genistein in 60 post-menopausal women for six months. This study showed that 54mg of genistein on daily dose showed therapeutic effect on both cardioprotective activity and glycemic control (Crisafulli et al., 2005). Deodato and colleagues used a rat model of coronary artery occlusion showed that an intravenous injection of 1 mg/ kg genistein after five minutes reduced myocardial necrosis, both in terms of necrotic area and serum and macrophage TNF-α levels, and blunted intercellular adhesion molecule-1 (ICAM-1) expression in the injured myocardium, as well as serum creatinine phosphokinase (CPK) activity, ventricular arrhythmias are less common, and myocardial contractility is better (Deodato et al., 1999). Thus genistein seems to protect against myocardial ischemia and reperfusion injury. It is demonstrated from genistein may have potential in maintaining health of heart in post-menopausal women.

9.3. Role of Genistein in Obesity

In post-menopause, due to decrease in estrogen level, alteration of lipid and carbohydrate metabolism occurs. It leads to increase in fat content and accumulation of fatty content in human body. In another study, it was found that estrogen induces the metabolic rate of lipids and carbohydrates, which decrease the body weight (Szkudelska et al., 2007). In post-menopausal women, administration of phytoestrogens reduces markers of obesity, such as Low-density Lipoprotein (LDL) cholesterol and total cholesterol (Jayagopal et al., 2002). It is mentioned that these markers may present in high concentration during obesity. Furthermore, genistein in adipocytes, enhance lipolysis as well it restricted lipogenesis and regulate adipose tissues

(Jayagopal et al., 2002; Szkudelska et al., 2007). Moreover, it was also shown that administration of genistein for 6-12 weeks in blood increases the level of high-density lipoprotein cholesterol of postmenopausal women having metabolic syndrome and healthy postmenopausal women and by decreasing the obesity risks (Szkudelska et al., 2007; Bitto et al., 2010). It was shown that genistein administration in ovariectomized rats responded in term of reduced body weight after two weeks of treatment (Thangavel et al., 2019). The effects of genistein on metabolic regulation of lipids and fats could be useful for women having obesity to improve the quality of life due to surgical and natural post-menopause.

9.4. Role of Genistein in Diabetes

According to World Health Organization (WHO) information, around 442 million grown-ups worldwide have diabetes mellitus (DM) that is about to one in each eleven individuals. DM causes confusions overall wellbeing and expands the danger of unexpected passing. It was assessed that roughly 1.5 million individuals bite the dust from DM. It ought to be noticed that ladies are more vulnerable to DM after normal or careful menopause because of the awkwardness or nonattendance of chemicals (Thangavel et al., 2019). The shortfall of ovarian chemicals after menopause or ovariectomy changes glucose digestion, subsequently prompting vacillations in glucose levels. Restricted information is accessible with respect to whether hormonal changes are connected to hindrances in glucose digestion and insulin affectability; a few, yet not all, ladies experienced modified glucose digestion and expanded danger of creating DM. HRT may have impartial or conceivably advantageous consequences for glucose digestion. Not-withstanding, HRT has been positive related with expanded danger of cardiovascular illness (CV), in this way its expected use for DM in postmenopausal ladies is dubious. Strangely, genistein has been displayed to manage this digestion, with ideal impacts being noticed (Thangavel et al., 2019). For instance, treatment with phytoestrogens in postmenopausal ladies with type 2 DM diminished fasting insulin and insulin obstruction following 12 weeks of treatment (Thangavel et al., 2019). The organization of genistein (54 mg/day) for 6, 12, or two years decreased the centralization of fasting glucose. These examinations likewise tracked down that expanded plasma genistein fixations further developed glucose homeostasis in the blood. Scarcely any examinations revealed hostile to diabetic impacts of genistein in hepatic female mice, in this way bringing about the decrease of blood glucose, glucagon, and HbA1c levels. Moreover, the exercises of proteins that used hepatic glucose were diminished by genistein; decreased in glucose-6-phosphatase (G6Pase) and phosphoenol pyruvate carboxykinase (PEPCK) were distinguished after genistein treatment (0.2 g/kg diet) for nine weeks in female mice, demonstrating that genistein upgraded glucose digestion (García-Jiménez et al., 1994; Thangavel et al., 2019). In addition, the organization of genistein (0.2 g/kg diet) in female mice for nine weeks expanded the action of lipogenic catalysts (malic compound and glucose-6-phosphate dehydrogenase). It was seen that the decrease in the action of lipogenic compounds adjusted glucose levels. Genistein followed up on these proteins and expanded movement, accordingly decreasing the danger of DM. These useful impacts of genistein in the guideline of glucose digestion show its latent capacity with respect to the enhancement of raised glucose levels, which prompts the improvement of DM in postmenopausal lady (Khandelwal and Pugazhenthi, 1995).

9.5. Antidepressant and Anxiolytic Effects of Genistein

As referenced above, normal as well as careful menopause prompts less or no creation of steroid chemicals in ladies. The decrease of estrogen and different chemicals in menopausal ladies can make depression and tension manifestations create. A few investigations have revealed the stimulant and anxiolytic impacts of genistein in the two people and trial creatures (Rodríguez-Landa et al., 2017). In certain examinations, genistein was demonstrated to work on personal satisfaction and burdensome manifestations in postmenopausal ladies. It was likewise shown that the microRNA miR221/222 assumed a significant part in the activity of energizer drugs (Baffa et al., 2010). Shen et al. (2018) contemplated the impact of genistein on miR221/222 in U87-MG cells brooded with 10 μM of genistein for 72 h. They found that genistein diminished/hindered the articulation level of miR221/222, however expanded the articulation level of Cx43 in contrast with the control, subsequently uncovering the upper impact of genistein. The increment in Cx43 levels in the cerebrum identified with a gentle type of pressure and melancholy. These outcomes were additionally predictable with in vivo tries, which demonstrated the capacity of genistein to smother some biomarkers related with significant sorrow (Shen et al., 2018). In addition, treatment with genistein (10 mg/kg) alone or in blend with amitriptyline (10 mg/kg) for 10 days delivered comparative outcomes to that of amitriptyline.

Albeit the system by which genistein applies its impact stays hazy, it was theorized that genistein may deliver its results because of its underlying likeness with estrogen. Moreover, genistein can infiltrate the blood–mind hindrance and tie differentially to ER-α or ER-β in the hippocampus of the cerebrum (Gupta et al., 2015), which is related with energizer impacts delivered by estrogens. A few examinations have detailed that energizer impacts include ER-β instead of ER-α. Also, genistein may manage the serotonergic pathway under unpleasant conditions by diminishing the serotonin turnover of 5-hydroxyindoleacetic corrosive/serotonin (5-HIAA/5-HT) proportion, demonstrating that the convergence of 5-HT in the hippocampus increments, as it happens with stimulant medications (Thangavel et al., 2019). The actuation of monoamine oxidase (MAO), which is liable for the digestion of serotonin and dopamine, is related with wretchedness side effects. The digestion of serotonin prompts a seriously hopeless mind-set and wretchedness. Organization of genistein diminishes melancholy manifestations by controlling MAO action in some cerebrum structures, which is related with energizer like impacts. Further examinations, like corticosterone estimations, pharmacokinetic studies, and toxicology appraisals, may show the possible adequacy of genistein as a stimulant specialist (Evans et al., 2011).

10. Current Status of Genistein

Authors summarized the accessible proof on the promising job of genistein against malignancy. A few exploratory and clinical examinations recommend a helpful job of genistein on various kinds of malignancy. The rise of negative phenomenon in cancer treatment is notable, for example, drug opposition, high danger of backslide, and the inaccessibility or helpless result of treatments, like a medical procedure, chemotherapy, phototherapy, and radiotherapy. In this way, consideration has been paid lately to normal cures having the ability to work on the adequacy of chemotherapeutic treatment and to bring down unfriendly impacts. Genistein can be incorporated among these mixtures in light of the fact that the atom shows synergistic conduct when related with notable anticancer medications, for example, adriamycin, tamoxifen and docetaxel, recommending an expected job in blend treatment. In any case, genistein, just as other bioactive phytochemicals, benefits and, simultaneously, experiences 2 obviously inverse highlights: high pleiotropy and low bioavailability (Tuli et al., 2021). The previous alludes to the capacity of an offered compound to act at a few levels in the cells, setting off simultaneously a few biochemical pathways associated with the event and improvement of disease (i.e., cell cycle capture, apoptosis, cell passing). The net outcome is a synergistic impact that may improve the adequacy of a particular medication, regardless of whether present in the cells at moderately low focuses. In such manner, in the past passages, we evaluated a few sub-atomic focuses of genistein, like ER, tyrosine kinases, and supportive of and antiapoptotic factors. Bioavailability was additionally talked about above, which carries us to the idea that "what we adsorb" from food is considerably more significant than what we eat: the plasma grouping of genistein present in the eating routine (like other bioactive mixtures) is fundamentally lower than the focuses required in trial models (cell lines and creature contemplates) to trigger an anticancer reaction (Tuli et al., 2021). Subsequently, it is sensible to anticipate a huge clinical result of genistein when applied at pharmacologic dosages (many micromolars) and frail or invalid impacts when a similar particle is directed at chemo preventive portions. The fact of the matter is most likely more mind boggling than we can anticipate. As a matter of fact, when genistein is adsorbed at low focuses along with other bioactive mixtures present in the eating routine, we can hypothesize pleiotropic anticancer impacts that outcome from synergistic systems owing to the plenty of individual mixtures (or their metabolites) getting from the eating regimen. On the other hand, we can likewise estimate that genistein has, at low dosages, impacts not quite the same as those deliberate at high portions, contingent upon the cell foundation and the atomic objective explored. In this regard, the capacity of genistein to restrain cell development in both chemical ward and - free malignant growth cells are portion subordinate (Russo et al., 2010). Truth be told, when genistein fixation increments from a low nanomolar focus to many nanomolars, special enactment of ER- β is lost and genistein actuates the two ERs (α and β); subsequently, basically in various breast malignant growths, the proportions of the ER subtypes and the concentrations of genistein unequivocally impact the last impact on cell destiny and hormone - regulated gene expression (Chang et al., 2008).

11. Future Perspectives of Genistein

The current study conducted a thorough analysis of the scientific literature on genistein's potential role as an anticancer drug, including a full description of its targets in signalling pathways. The chemopreventive potential of genistein was further highlighted in *in vivo* and *in vitro* research, prompting genistein to be up scaled to clinical trials. This is especially promising for patients in developing nations, where currently available cancer treatments are prohibitively expensive and out of reach for most patients, restricting therapy options. As a result, developing nations are frantically seeking low-cost cancer prevention and treatment solutions, and organic products as anticancer drugs appear to be promising. This highlights the importance of developing safe and effective natural product-derived chemopreventive medicines that are readily available. Given genistein's significance in a wide range of molecular pathways, there is a growing need to comprehend this important chemical in terms of both its regulatory and therapeutic potential. The experimental and clinical evidence presented throughout this review points to genistein's potential therapeutic function in the prevention and treatment of several cancers, and it encourages more in vivo research with this appealing natural chemical. As a result, synergistic techniques utilising genistein may soon aid in the treatment of severe disorders such as cancer.

Although cancer therapy modalities have progressed significantly, there are still untapped dimensions to scale for adjusting effective treatment techniques for cancer patients with long-term safety and efficacy. Natural compounds like genistein have been investigated extensively for their anticancer properties, which are detailed in this article. However, we seem to have a far toward to go, though. First, in addition to the recognized biomarkers as therapeutic modulators, more clinical trials on large patient cohorts are needed to thoroughly investigate genistein's chemopreventive and therapeutic potential, as well as its long-term safety and effectiveness in the treatment of cancer patients. Second, the mechanisms and tactics that could be used to increase the accessibility and bioavailability of genistein should be investigated further. Finally, the exact therapeutic dose of genistein for the therapy of various cancer types must be standardised. Fourth, based on different effects reported in long-term and short-term therapy administration regimens, the duration and time of genistein dose administration require extensive standardisation. Similarly, the use of nanotechnology to minimise the required doses of genistein and to specifically target tumour regions utilising innovative drug delivery strategies should be investigated in the future. As a result, considerable preclinical research as well as clinical studies must be planned and carried out before genistein becomes a popular cancer treatment option. Similarly, nanotechnology should be researched in the future to reduce the required doses of genistein and to specifically target tumour locations using new drug delivery systems. As a result, extensive clinical and preclinical investigation must be planned and carried out before genistein becomes a widely used cancer treatment (Tuli et al., 2019).

Genistein works as a chemotherapeutic agent by affecting a variety of processes, including apoptosis, cell cycle, and angiogenesis, as well as blocking metastasis (Gu et al., 2005; Thangavel et al., 2019). The key targets include kinesin-like protein 20A (KIF20A), Wnt/β-catenin, PI3K/AKT, ERK1/2, caspases, Bax, Bcl-2, NF-κB, and MAPK, as indicated above, and signalling pathways may operate as the anti-inflammatory molecular mechanisms; among others, genistein's anticancer therapeutic effects (Zhou et al., 2001a,b; Yang et al., 2010; Mansoor et al., 2011; Wang et al., 2014, Du et al., 2018; Thangavel et al., 2019). Genistein's synergistic behaviour with well-known anticancer medications including adriamycin, docetaxel, and tamoxifen suggested that it could play a role in synergistic therapy (Chen et al., 2005). Despite the facts, that genistein appears to have a wide range of anticancer properties, its effectiveness is debatable. Studies on the link between soy consumption and the risk of breast cancer are contradictory, because foods high in phytoestrogens, such as genistein, may have complicated actions exerting both promoting and preventative effects, soy consumption and breast cancer risk are at odds (Guo et al., 2019). As previously stated, greater genistein consumption is linked to a lower risk of breast, lung, colorectal, ovarian, and endometrial cancers (Reger et al., 2018). Natural chemicals are currently being investigated in a therapeutic environment to modify molecular pathways, which could lead to increased control over treatment regimens in the future. Furthermore, pharmaceutical targets for cancer therapy necessitate a deeper understanding of the interactions and mechanisms involved. More research into genistein's possible clinical applications will be conducted and will assist in resolving the difficulties surrounding the usage of genistein (Thangavel et al., 2019).

Conclusion

The poor outcomes of existing medicines and associated side effects cannot be eradicated, and negative occurrences in the treatment of numerous postmenopausal symptoms are well-known. The use of genistein to treat postmenopausal symptoms has shown to be effective, with few to no negative side effects. However, much more complementary study and insights on the molecular level are needed in this field, and more research into the molecular pathways via which genistein works will help to enhance new treatments for postmenopausal symptoms and other disorders. Even though the biological effects are more promising having certain restrictions, such as effects on target organs, low bioavailability and biological estrogenic activity, but have limited the clinical bids of genistein to some extent. Besides, few studies reported that molecular structure modification of genistein may eliminate the activity of estrogen and target organs. In the present review, we summarize the latent benefits of genistein on symptoms of menopause and disease-related to menopause like osteoporosis, anxiety, obesity, cardiovascular, diabetes, breast cancer and depression.

References

Ahmad, S. & Pathak, D. (2000). Nutritional changes in soybean during germination. *Journal of Food Science and Technology*, 37, 665–666.

Arai, N., Ström, A., Rafter, J. J. & Gustafsson, J. A. (2000). Estrogen receptor beta mRNA in colon cancer cells: Growth effects of estrogen and genistein. *Biochemical and Biophysical Research Communications*, 270(2), 425–431.

Atteritano, M., Marini, H., Minutoli, L., Polito, F., Bitto, A., Altavilla, D., Mazzaferro, S., D'Anna, R., Cannata, M. L., Gaudio, A., Frisina, A., Frisina, N., Corrado, F., Cancellieri, F., Lubrano, C., Bonaiuto, M., Adamo, E. B. & Squadrito, F. (2007). Effects of the phytoestrogen genistein on some predictors of cardiovascular risk in osteopenic, postmenopausal women: A two-year randomized, double-blind, placebo-controlled study. *Journal of Clinical Endocrinology and Metabolism*, 92(8), 3068–3075.

Baffa, A., Hohoff, C., Baune, B. T., Müller-Tidow, C., Tidow, N., Freitag, C., Zwanzger, P., Deckert, J., Arolt, V. & Domschke, K. (2010). Norepinephrine and serotonin transporter genes: Impact on treatment response in depression. *Neuropsychobiology*, 62(2), 121–131.

Baker, W. & Robinson, R. (1928). Synthetical experiments in the iso flavone group. Part III. A synthesis of genistein. *Journal of the Chemical Society*, 3115–3118.

Barnes, S., Kim, H. & Usmar, V. D. (2000). Beyond ERα and ERβ: Estrogen receptor binding is only part of the isoflavone story. *Journal of Nutrition*, 130, 656S–657S.

Barratt, G. (2003). Colloidal drug carriers: Achievements and perspectives. *Cellular and Molecular Life Sciences*, 60(1), 21–37.

Bhagwat, S., Haytowitz, D. B. & Holden, J. M. (2008). *USDA database for the isoflavone content of selected foods. Release 2.0.*

Bitto, A., Polito, F., Atteritano, M., Altavilla, D., Mazzaferro, S., Marini, H., Adamo, E. B., D'Anna, R., Granese, R., Corrado, F., Russo, S., Minutoli, L. & Squadrito, F. (2010). Genistein aglycone does not affect thyroid function: Results from a three-year, randomized, double-blind, placebo-controlled trial. *Journal of Clinical Endocrinology and Metabolism*, 95(6), 3067–3072.

Caretto, M., Giannini, A., Simoncini, T. & Genazzani, A. R. (2018). Menopause and ageing. In J. Schenker, J. Sciarra, L. Mettler, A. Genazzani & M. Birkhaeuser (Eds.), *Reproductive Medicine for Clinical Practice. Springer,* pp. 177–189.

Cassidy, A. (1999). Potential tissue selectivity of dietary phytoestrogens and estrogens. *Current Opinion in Lipidology*, 10(1), 47–52.

Chan, K. K. L., Siu, M. K. Y., Jiang, Y. X., Wang, J. J., Leung, T. H. Y. & Ngan, H. Y. S. (2018). Estrogen receptor modulators genistein, daidzein and ERB-041 inhibit cell migration, invasion, proliferation and sphere formation via modulation of FAK and PI3K/ AKT signaling in ovarian cancer. *Cancer Cell International*, 18, 65. https://doi.org/10.1186/s12935-018-0559-2.

Chandrasekharan, S. & Aglin, A. (2013). Pharmacokinetics of dietary isoflavones. *J. Steroids Hormon Sci. Suppl.*, s12, 2–8.

Chang, E. C., Charn, T. H., Park, S. H., Helferich, W. G., Komm, B., Katzenellenbogen, J. A. & Katzenellenbogen, B. S. (2008). Estrogen receptors alpha and beta as determinants of gene expression: Influence of ligand, dose, and chromatin binding. *Molecular Endocrinology*, 22(5), 1032–1043.

Chang, H. C. & Doerge, D. R. (2000). Dietary genistein inactivates rat thyroidperoxidase *in vivo* without an apparent hypothyroid effect. *Toxicology and Applied Pharmacology*, 168(3), 244–252.

Chang, Y. C., Nair, M. G., Santell, R. C. & Helferich, W. G. (1994). Microwave-mediated synthesis of anticarcinogenic isoflavones from soybeans. *Journal of Agricultural and Food Chemistry*, 42(9), 1869–1871.

Chatzinikolaou, G., Nikitovic, D., Stathopoulos, E. N., Velegrakis, G. A., Karamanos, N. K. & Tzanakakis, G. N. (2007). Protein tyrosine kinase and estrogen receptor-dependent pathways regulate the synthesis and distribution of glycosamino-glycans/proteoglycans produced by two human colon cancer cell lines. *Anticancer Research*, 27(6B), 4101–4106.

Chen, W. F., Huang, M. H., Tzang, C. H., Yang, M. & Wong, M. S. (2003). Inhibitory actions of genistein in human breast cancer (MCF-7) cells. *Biochimica et Biophysica Acta*, 1638(2), 187–196.

Chen, Y., Wang, Y., Xu, W. & Zhang, Z. (2005). Analysis on the mechanism of Helicobacter pylori-induced apoptosis in gastric cancer cell line BGC-823. *International Journal of Molecular Medicine*, 16(4), 741–745.

Chodon, D., Ramamurty, N. & Sakthisekaran, D. (2007). Preliminary studies on induction of apoptosis by genistein on HepG2 cell line. *Toxicology in Vitro*, 21(5), 887–891.

Choi, J. N., Kim, D., Choi, H. K., Yoo, K. M., Kim, J. & Lee, C. H. (2009). 29Hydroxylation of genistein enhanced antioxidant and antiproliferative activities in Mcf-7 human breast cancer cells. *Journal of Microbiology and Biotechnology*, 19, 1348–1354.

Cotterchio, M., Boucher, B. A., Manno, M., Gallinger, S., Okey, A. & Harper, P. (2006). Dietary phytoestrogen intake is associated with reduced colorectal cancer risk. *Journal of Nutrition*, 136(12), 3046–3053.

Crisafulli, A., Altavilla, D., Marini, H., Bitto, A., Cucinotta, D., Frisina, N., Corrado, F., D'Anna, R., Squadrito, G., Adamo, E. B., Marini, R., Romeo, A., Cancellieri, F.,

Buemi, M. & Squadrito, F. (2005). Effects of the phytoestrogen genistein on cardiovascular risk factors in postmenopausal women. *Menopause*, 12(2), 186–192.

de Oliveira, M. R. (2016). Evidence for genistein as a mitochondriotropic molecule. *Mitochondrion*, 29, 35–44.

de Oliveira, S. R., Taveira, S. F., Marreto, R. N., Valadares, M. C., Diniz, D. G. A. & Lima, E. M. (2013). Preparation and characterization of solid oral dosage forms containing soy isoflavones. *Revista Brasileira de Farmacognosia*, 23(1), 175–181.

Deodato, B., Altavilla, D., Squadrito, G., Campo, G. M., Arlotta, M., Minutoli, L., Saitta, A., Cucinotta, D., Calapai, G., Caputi, A. P., Miano, M. & Squadrito, F. (1999). Cardioprotection by the phytoestrogen genistein in experimental myocardial ischaemia-reperfusion injury. *British Journal of Pharmacology*, 128(8), 1683–1690.

Desai, J. & Thakkar, H. (2016). Effect of particle size on oral bioavailability of Darunavir-loaded solid lipid nanoparticles. *Journal of Microencapsulation*, 33(7), 669–678.

Dixon, R. A. & Ferreira, D. (2002). Genistein. *Phytochemistry*, 60(3), 205–211.

Du, Z. R., Feng, X. Q., Li, N., Qu, J. X., Feng, L., Chen, L. & Chen, W. F. (2018). G protein-coupled estrogen receptor is involved in the anti-inflammatory effects of genistein in microglia. *Phytomedicine*, 43, 11–20.

El-Serag, H. B., Marrero, J. A., Rudolph, L. & Reddy, K. R. (2008). Diagnosis and treatment of hepatocellular carcinoma. *Gastroenterology*, 134(6), 1752–1763.

Evans, B. A., Griffiths, K. & Morton, M. S. (1995). Inhibition of 5 alpha-reductase in genital skin fibroblasts and prostate tissue by dietary lignans and isoflavonoids. *Journal of Endocrinology*, 147(2), 295–302.

Evans, M., Elliott, J. G., Sharma, P., Berman, R. & Guthrie, N. (2011). The effect of synthetic genistein on menopause symptom management in healthy postmenopausal women: A multi-center, randomized, placebo-controlled study. *Maturitas*, 68(2), 189–196.

Fan, Y. E., Li, J., Liu, R., Ye, Z. G., Hu, J. N., Deng, Z. Y. & Peng, Y. (2011). Oral bioavailability of novel genistein sulfonates and their pre-clinical pharmacokinetics. *Latin American Journal of Pharmacy*, 30, 1582–1589.

Farina, H. G., Pomies, M., Alonso, D. F. & Gomez, D. E. (2006). Antitumor and antiangiogenic activity of soy isoflavone genistein in mouse models of melanoma and breast cancer. *Oncology Reports*, 16(4), 885–891.

Food Standards Agency. (2003). Phytoestrogens and health, committee on toxicology of chemicals in food, consumer products and the environment *https://cot.food.gov.uk/sites/default/files/cot/phytoreport0503.pdf*.

Fotsis, T., Pepper, M., Adlercreutz, H., Fleischmann, G., Hase, T., Montesano, R. & Schweigerer, L. (1993). Genistein, a dietary-derived inhibitor of in vitro angiogenesis. *Proceedings of the National Academic Science USA*, 90(7), 2690–2694.

Fu, X. H., Wang, L., Zhao, H., Xiang, H. L. & Cao, J. G. (2008). Synthesis of genistein derivatives and determination of their protective effects against vascular endothelial cell damages caused by hydrogen peroxide. *Bioorganic and Medicinal Chemistry Letters*, 18(2), 513–517.

Ganai, A. A. & Farooqi, H. (2015). Bioactivity of genistein: A review of in vitro and in vivo studies. *Biomedicine and Pharmacotherapy*, 76, 30–38.

García-Jiménez, C., Benito, B., Jolin, T. & Santisteban, P. (1994). Insulin regulation of malic enzyme gene expression in rat liver: Evidence for nuclear proteins that bind to two putative insulin response elements. *Molecular Endocrinology*, 8(10), 1361–1369.

Gartoulla, P., Bell, R. J., Worsley, R. & Davis, S. R. (2016). Menopausal vasomotor symptoms are associated with poor self-assessed work ability. *Maturitas*, 87, 33–39.

Goj, K., Rusin, A., Szeja, W., Kitel, R., Komor, R. & Grynkiewicz, G. (2012). Synthesis of genistein 2,3-Anhydroglycoconjugates ń potential antiproliferative agents. *Acta Poloniae Pharmaceutica*, 69(6), 1239–1247.

Gu, Y., Zhu, C. F., Iwamoto, H. & Chen, J. S. (2005). Genistein inhibits invasive potential of human hepatocellular carcinoma by altering cell cycle, apoptosis, and angiogenesis. *World Journal of Gastroenterology*, 11(41), 6512–6517.

Guo, P. P., Li, P., Zhang, X. H., Liu, N., Wang, J., Chen, D. D., Sun, W. J. & Zhang, W. (2019). Complementary and alternative medicine for natural and treatment-induced vasomotor symptoms: An overview of systematic reviews and meta-analyses. *Complementary Therapies in Clinical Practice*, 36, 181–194.

Gupta, G., Jia, T., Yee Woon, L., Kumar Chellappan, D., Candasamy, M. & Dua, K. (2015). Pharmacological evaluation of antidepressant-like effect of genistein and its combination with amitriptyline: An acute and chronic study. *Advances in Pharmacological Sciences*, 164943. https://doi.org/10.1155/2015/164943.

Hämäläinen, M., Nieminen, R., Vuorela, P., Heinonen, M. & Moilanen, E. (2007). Anti-inflammatory effects of flavonoids: Genistein, kaempferol, quercetin, and daidzein inhibit STAT-1 and NF-kappaB activations, whereas flavone, isorhamnetin, naringenin, and pelargonidin inhibit only NF-kappaB activation along with their inhibitory effect on iNOS expression and NO production in activated macrophages. *Mediators of Inflammation*, 45673. https://doi.org/10.1155/2007/45673.

Hara, A., Sasazuki, S., Inoue, M., Miura, T., Iwasaki, M., Sawada, N., Shimazu, T., Yamaji, T., Tsugane, S. & Japan Public Health Center-Based Prospective Study Group. (2013). Plasma isoflavone concentrations are not associated with gastric cancer risk among Japanese men and women. *Journal of Nutrition*, 143(8), 1293–1298.

Hong, X., Liu, P., Liang, J. & Yun, F. U. (2008). Genistein induces apoptosis through upregulation of p53 signaling pathway. *Journal of Tropical Medicine*, 9, 4.

Hörter, D. & Dressman, J. B. (2001). Influence of physicochemical properties on dissolution of drugs in the gastrointestinal tract. *Advanced Drug Delivery Reviews*, 46(1–3), 75–87.

Hsieh, C. Y., Santell, R. C., Haslam, S. Z. & Helferich, W. G. (1998). Estrogenic effects of genistein on the growth of estrogen receptor-positive human breast cancer (MCF-7) cells in vitro and in vivo. *Cancer Research*, 58(17), 3833–3838.

Huang, A. S., Hsieh, O. A. L. & Chang, S. S. (1982). Characterization of the non-volatile minor constituents responsible for the objectionable taste of defatted soybean flour. *Journal of Food Science*, 47(1), 19–23.

Huang, J., Nasr, M., Kim, Y. & Matthews, H. R. (1992). Genistein inhibits protein histidine kinase. *Journal of Biological Chemistry*, 267(22), 15511–15515.

Hwang, J. T., Lee, Y. K., Shin, J. I. & Park, O. J. (2009). Anti-inflammatory and anticarcinogenic effect of genistein alone or in combination with capsaicin in TPA-treated rat mammary glands or mammary cancer cell line. *Annals of the New York Academy of Sciences*, 1171, 415–420.

Hyz, K., Kawęcki, R., Misior, A., Bocian, W., Bednarek, E. B., Sitkowski, J. & Kozerski, L. (2011). Genistein binding mode to doubly nicked dumbbell DNA: Dynamic and diffusion ordered NMR study. *Journal of Medicinal Chemistry*, 54(24), 8386–8393.

Irace, C., Marini, H., Bitto, A., Altavilla, D., Polito, F., Adamo, E. B., Arcoraci, V., Minutoli, L., Di Benedetto, A., Di Vieste, G., de Gregorio, C., Gnasso, A., Corrao, S., Licata, G. & Squadrito, F. (2013). Genistein and endothelial function in postmenopausal women with metabolic syndrome. *European Journal of Clinical Investigation*, 43(10), 1025–1031.

Jaiswal, N., Akhtar, J., Singh, S. & Ahsan, F. (2018). An overview on genistein and its various formulations. *Drug Research*. https://doi.org/10.1055/a-0797-3657.

Jayagopal, V., Albertazzi, P., Kilpatrick, E. S., Howarth, E. M., Jennings, P. E., Hepburn, D. A. & Atkin, S. L. (2002). Beneficial effects of soy phytoestrogen intake in postmenopausal women with type 2 diabetes. *Diabetes Care*, 25(10), 1709–1714.

Jeong, J. W., Lee, H. H., Han, M. H., Kim, G. Y., Kim, W. J. & Choi, Y. H. (2014). Anti-inflammatory effects of genistein via suppression of the toll-like receptor 4-mediated signaling pathway in lipopolysaccharidestimulated BV2 microglia. *Chemico-Biological Interactions*, 212, 30–39.

Ji, G., Zhang, Y., Yang, Q., Cheng, S., Hao, J., Zhao, X. & Jiang, Z. (2012). Genistein suppresses LPS-induced inflammatory response through inhibiting NF-kappaB following AMP kinase activation in RAW 264.7 macrophages. *PLOS ONE*, 7(12), e53101.

Katsuyama, Y., Miyahisa, I., Funa, N. & Horinouchi, S. (2007). One-pot synthesis of genistein from tyrosine by coincubation of genetically engineered Escherichia coli and Saccharomyces cerevisiae cells. *Applied Microbiology and Biotechnology*, 73(5), 1143–1149.

Keating, E. & Martel, F. (2018). Antimetabolic effects of polyphenols in breast cancer cells: Focus on glucose uptake and metabolism. *Frontiers in Nutrition*, 5, 25. https://doi.org/10.3389/fnut.2018.00025

Kgomotso, T., Chiu, F. & Ng, K. (2008). Genistein-and daidzein 7-O-b-D-glucuronic acid retain the ability to inhibit copper-mediated lipid oxidation of low density lipoprotein. *Molecular Nutrition and Food Research*, 52(12), 1457–1466.

Khandelwal, R. L. & Pugazhenthi, S. (1995). In vivo effects of vanadate on hepatic glycogen metabolizing and lipogenic enzymes in insulin-dependent and insulin-resistant diabetic animals. *Molecular and Cellular Biochemistry*, 153(1–2), 87–94.

Kim, H., Peterson, T. G. & Barnes, S. (1998). Mechanisms of action of the soy isoflavone genistein: Emerging role for its effects via transforming growth factor b signaling pathways. *American Journal of Clinical Nutrition*, 68(6), 1418S–1425S.

Kim, J. T., Barua, S., Kim, H., Hong, S. C., Yoo, S. Y., Jeon, H., Cho, Y., Gil, S., Oh, K. & Lee, J. (2017). Absorption study of genistein using solid lipid microparticles and nanoparticles: Control of oral bioavailability by particle sizes. *Biomolecules and Therapeutics*, 25(4), 452–459.

Kim, S. H., Kim, C. W., Jeon, S. Y., Go, R. E., Hwang, K. A. & Choi, K. C. (2014). Chemopreventive and chemotherapeutic effects of genistein, a soy isoflavone, upon cancer development and progression in preclinical animal models. *Laboratory Animal Research*, 30(4), 143–150.

Kim, W. J., Lee, H. Y., Won, M. H. & Yoo, S. H. (2005). Germination effect of soybean on its contents of isoflavones and oligosaccharides. *Food Science and Biotechnology*, 14, 498–502.

Kiriakidis, S., Högemeier, O., Starcke, S., Dombrowski, F., Hahne, J. C., Pepper, M., Jha, H. C. & Wernert, N. (2005). Novel tempeh (fermented soyabean) isoflavones inhibit in vivo angiogenesis in the chicken chorioallantoic membrane assay. *British Journal of Nutrition*, 93(3), 317–323.

Klinge, C. M. (2000). Estrogen receptor interaction with co-activators and corepressors. *Steroids*, 65(5), 227–251.

Ko, K. P., Park, S. K., Park, B., Yang, J. J., Cho, L. Y., Kang, C., Kim, C. S., Gwack, J., Shin, A., Kim, Y., Kim, J., Yang, H. K., Kang, D., Chang, S. H., Shin, H. R. & Yoo, K. Y. (2010). Isoflavones from phytoestrogens and gastric cancer risk: A nested case-control study within the Korean Multicenter Cancer Cohort. *Cancer Epidemiology, Biomarkers and Prevention*, 19(5), 1292–1300.

Kochs, G. & Grisebach, H. (1986). Enzymic synthesis of isoflavones. *European Journal of Biochemistry*, 155(2), 311–318.

Kuiper, G. G., Lemmen, J. G., Carlsson, B., Corton, J. C., Safe, S. H., van der Saag, P. T., van der Burg, B. & Gustafsson, J. A. (1998). Interaction of estrogenic chemicals and phytoestrogens with estrogen receptor beta. *Endocrinology*, 139(10), 4252–4263.

Kurzer, M. S. (2002). Hormonal effects of soy in premenopausal women and men. *Journal of Nutrition*, 132(3), 570S–573S.

Kurzer, M. S. (2003). Phytoestrogen supplement use by women. *Journal of Nutrition*, 133(6), 1983S–1986S.

Kwon, S. H., Kang, M. J., Huh, J. S., Ha, K. W., Lee, J. R., Lee, S. K., Lee, B. S., Han, I. H., Lee, M. S., Lee, M. W., Lee, J. & Choi, Y. W. (2007). Comparison of oral bioavailability of genistein and genistin in rats. *International Journal of Pharmaceutics*, 337(1–2), 148–154.

Lee, S. R., Kwon, S. W., Lee, Y. H., Kaya, P., Kim, J. M., Ahn, C., Jung, E. M., Lee, G. S., An, B. S., Jeung, E. B., Park, B. K. & Hong, E. J. (2019). Dietary intake of genistein suppresses hepatocellular carcinoma through AMPK-mediated apoptosis and anti-inflammation. *BMC Cancer*, 19(1), 6.

Lee, S. Y., Lee, S., Lee, S., Oh, J. Y., Jeon, E. J., Ryu, H. S. & Lee, C. H. (2014). Primary andsecondary metabolite profiling of doenjang, a fermented soybean pasteduring industrial processing. *Food Chemistry*, 165, 157–166.

Li, J., Li, J., Yue, Y., Hu, Y., Cheng, W., Liu, R., Pan, X. & Zhang, P. (2015). Genistein suppresses tumor necrosis factor alpha-induced inflammation via modulating reactive oxygen species/Akt/nuclear factor kappa B and adenosine monophosphateactivated protein kinase signal pathways in human synoviocyte MH7A cells. *Drug Design, Development and Therapy*, 8, 315–323.

Li, S., Li, J., Dai, W., Zhang, Q., Feng, J., Wu, L., Liu, T., Yu, Q., Xu, S., Wang, W., Lu, X., Chen, K., Xia, Y., Lu, J., Zhou, Y., Fan, X., Mo, W., Xu, L. & Guo, C. (2017). Genistein suppresses aerobic glycolysis and induces hepatocellular carcinoma cell death. *British Journal of Cancer*, 117(10), 1518–1528.

Li, W., Jia, H. Y., He, X. H., Shi, W. G. & Zhong, B. H. (2012). Novel phenoxyalkylcarboxylic acid derivatives as hypolipidaemic agents. *Journal of Enzyme Inhibition and Medicinal Chemistry*, 27(2), 311–318.

Liggins, J., Bluck, L. J., Runswick, S., Atkinson, C., Coward, W. A. & Bingham, S. A. (2000a). Daidzein and genistein contents of vegetables. *British Journal of Nutrition*, 84(5), 717–725.

Liggins, J., Bluck, L. J., Runswick, S., Atkinson, C., Coward, W. A. & Bingham, S. A. (2000b). Daidzein and genistein content of fruits and nuts. *Journal of Nutritional Biochemistry*, 11(6), 326–331.

Liu, H. & Yu, G. Y. (2004). Antimetastatic effects of genistein on salivary adenoid cystic carcinoma in vivo.*Zhonghua Kou Qiang Yi Xue Za Zhi*, 39(5), 373–375.

Magee, P. J. & Rowland, I. R. (2004). Phyto-oestrogens, their mechanism of action: Current evidence for a role in breast and prostate cancer. *British Journal of Nutrition*, 91(4), 513–531.

Maniewska, J., Grynkiewicz, G., Szeja, W. & Hendrich, A. B. (2009). Interaction of genistein benzyl derivatives with lipid bilayers-fluorescence spectroscopic and calorimetric study. *Bioorganic and Medicinal Chemistry*, 17(6), 2592–2597.

Mansoor, T. A., Ramalho, R. M., Luo, X., Ramalhete, C., Rodrigues, C. M. & Ferreira, M. J. (2011). Isoflavones as apoptosis inducers in human hepatoma HuH-7 cells. *Phytotherapy Research*, 25(12), 1819–1824.

Mareti, E., Abatzi, C., Vavilis, D., Lambrinoudaki, I. & Goulis, D. G. (2019). Effect of oral phytoestrogens on endometrial thickness and breast density of perimenopausal and postmenopausal women: A systematic review and meta-analysis. *Maturitas*, 124, 81–88.

Marini, H., Minutoli, L., Polito, F., Bitto, A., Altavilla, D., Atteritano, M., Gaudio, A., Mazzaferro, S., Frisina, A., Frisina, N., Lubrano, C., Bonaiuto, M., D'Anna, R., Cannata, M. L., Corrado, F., Adamo, E. B., Wilson, S. & Squadrito, F. (2007). Effects of the phytoestrogen genistein on bone metabolism in osteopenic postmenopausal women: A randomized trial. *Annals of Internal Medicine*, 146(12), 839–847.

Markovits, J., Linassier, C., Fossé, P., Couprie, J., Pierre, J., Jacquemin-Sablon, A., Saucier, J. M., Le Pecq, J. B. & Larsen, A. K. (1989). Inhibitory effects of the tyrosine kinase inhibitor genistein on mammalian DNA topoisomerase II. *Cancer Research*, 49(18), 5111–5117.

Matsuda, H., Shimoda, H., Morikawa, T. & Yoshikawa, M. (2001). Phytoestrogens from the roots of Polygonum cuspidatum (Polygonaceae): Structure-requirement of hydroxyl anthraquinones for estrogenic activity. *Bioorganic and Medicinal Chemistry Letters*, 11(14), 1839–1842.

Matsumoto, T., Kobayashi, T., Kikuchi, T., Honda, T. & Kamata, K. (2005). Effects of dual-action genistein derivatives on relaxation in rat aorta. *Journal of Smooth Muscle Research*, 41(1), 23–33.

Mazumder, R. M. A. & Hongsprabhas, P. (2016). Genistein as antioxidant and antibrowning agents in vivo and in vitro: A review. *Biomedicine and Pharmacotherapy*, 82, 379–392.

Meng, Q. H., Wähälä, K., Adlercreutz, H. & Tikkanen, M. J. (1999). Antiproliferative efficacy of lipophilic soy isoflavone phytoestrogens delivered by low density lipoprotein particles into cultured U937 cells. *Life Sciences*, 65(16), 1695–1705.

Messina, M., Nagata, C. & Wu, A. H. (2006). Estimated Asian adult soy protein and isoflavone intakes. *Nutrition and Cancer*, 55(1), 1–12.

Michikawa, T., Inoue, M., Sawada, N., Tanaka, Y., Yamaji, T., Iwasaki, M., Shimazu, T., Sasazuki, S., Mizokami, M., Tsugane, S. & Japan Public Health Center-based Prospective Study Group. (2015). Plasma isoflavones and risk of primary liver cancer in Japanese women and men with hepatitis virus infection: A nested case-control study. *Cancer Epidemiology, Biomarkers and Prevention*, 24(3), 532–537.

Miękus, K. & Madeja, Z. (2007). Genistein inhibits the contact-stimulated migration of prostate cancer cells. *Cellular and Molecular Biology Letters*, 12(3), 348–361.

Milligan, S. R., Kalita, J. C., Heyerick, A., Rong, H., De Cooman, L. & De Keukeleire, D. (1999). Identification of a potent phytoestrogen in hops (*Humulus lupulus* L.) and beer. *Journal of Clinical Endocrinology and Metabolism*, 84(6), 2249–2252.

Mohanraj, V. J. & Chen, Y. (2006). Nanoparticles- A review. *Tropical Journal of Pharmaceutical Research*, 5(1), 561–573.

Motlekar, N., Khan, M. A. & Youan, B. C. (2006). Preparation and characterization of genistein containing poly (ethylene glycol) microparticles. *Journal of Applied Polymer Science*, 101(3), 2070–2078.

Mukherjee, S., Acharya, B. R., Bhattacharyya, B. & Chakrabarti, G. (2010). Genistein arrests cell cycle progression of A549 cells at the G (2). *Biochemistry*, 49(8), 1702–1712.

Nguyen, D. T., Hernandez-Montes, E., Vauzour, D., Schönthal, A. H., Rice-Evans, C., Cadenas, E. & Spencer, J. P. E. (2006). The intracellular genistein metabolite 5, 7, 39, 49-tetrahydroxyisoflavone mediates G2-M cell cycle arrest in cancer cells via modulation of the p38 signaling pathway. *Free Radical Biology and Medicine*, 41(8), 1225–1239.

Nogowski, L., Nowak, K. W., Kaczmarek, P. & Maćkowiak, P. (2002). The influence of coumestrol, zearalenone, and genistein administration on insulin receptors and insulin secretion in ovariectomized rats. *Journal of Receptor and Signal Transduction Research*, 22(1–4), 449–457.

Ohno, S., Nakajima, Y., Inoue, K., Nakazawa, H. & Nakajin, S. (2003). Genistein administration decreases serum corticosterone and testosterone levels in rats. *Life Sciences*, 74(6), 733–742.

Ohno, S., Shinoda, S., Toyoshima, S., Nakazawa, H., Makino, T. & Nakajin, S. (2002). Effects of flavonoid phytochemicals on cortisol production and on activities of steroidogenic enzymes in human adrenocortical H295R cells. *Journal of Steroid Biochemistry and Molecular Biology*, 80(3), 355–363.

Ososki, A. L. & Kennelly, E. J. (2003). Phytoestrogens: A review of the present state of research. *Phytotherapy Research*, 17(8), 845–869.

Paucar-Menacho, L. M., Berhow, M. A., Mandarino, J. M. G., Chang, Y. K. & Mejia, E. Gd. (2010). Effect of time and temperature on bioactive compounds in germinated Brazilian soybean cultivar BRS 258. *Food Research International*, 43(7), 1856–1865.

Perkin, A. G. & Newbury, F. G. (1899). LXXIX.-The colouring matters contained in dyer's broom (*Genista tinctoria*) and heather (*Calluna vulgaris*). *Journal of the Chemical Society Transactions*, 75, 830–839.

Pintova, S., Dharmupari, S., Moshier, E., Zubizarreta, N., Ang, C. & Holcombe, R. F. (2019). Genistein combined with FOLFOX or FOLFOX-bevacizumab for the treatment of metastatic colorectal cancer: phase I/II pilot study. *Cancer Chemotherapy and Pharmacology*, 84(3), 591–598.

Polkowski, K. & Mazurek, A. P. (2000). Biological properties of genistein: A review of in vitro and in vivo data. *Acta Poloniae Pharmaceutica*, 57(2), 135–155.

Pool, H., Campos-Vega, R., Herrera-Hernández, M. G., García-Solis, P., García-Gasca, T., Sánchez, I. C., Luna-Bárcenas, G. & Vergara-Castañeda, H. (2018). Development of genistein-pegylated silica hybrid nanomaterials with enhanced antioxidant and antiproliferative properties o HT29 human colon cancer cells. *American Journal of Translational Research*, 10(8), 2306–2323.

Potter, S. M. (1995). Overview of proposed mechanisms for the hypocholesterolemic effect of soy. *Journal of Nutrition*, 125(3), 606S–611S.

Qi, W., Weber, C. R., Wasland, K. & Savkovic, S. D. (2011). Genistein inhibits proliferation of colon cancer cells by attenuating a negative effect of epidermal growth factor on tumor suppressor FOXO3 activity. *BMC Cancer*, 11, 219. https://doi.org/10.1186/1471-2407-11-219.

Qiang, X., Sang, Z., Yuan, W., Li, Y., Liu, Q., Bai, P., Shi, Y., Ang, W., Tan, Z. & Deng, Y. (2014). Design, synthesis and evaluation of genistein-O-alkylbenzylamines as potential multifunctional agents for the treatment of Alzheimer's disease. *European Journal of Medicinal Chemistry*, 76, 314–331.

Quinhone, A., Jr. & Ida, E. I. (2015). Profile of the contents of different forms of soybean isoflavones and the effect of germination time on these compounds and the physical parameters in soybean sprouts. *Food Chemistry*, 166, 173–178.

Rafi, M. M., Rosen, R. T., Vassil, A., Ho, C. T., Zhang, H., Ghai, G., Lambert, G. & DiPaola, R. S. (2000). Modulation of bcl-2 and cytotoxicity by licochalcone-A, a novel estrogenic flavonoid. *Anticancer Research*, 20(4), 2653–2658.

Ravindranath, M. H., Muthugounder, S., Presser, N. & Viswanathan, S. (2004). Anticancer therapeutic potential of soy isoflavone, genistein. *In:* Cooper E.L., Yamaguchi N. (eds) Complementary and alternative approaches to biomedicine. *Advances in Experimental Medicine and Biology*, Springer, 546. https://doi.org/10.1007/978-1-4757-4820-8_11.

Reger, M. K., Zollinger, T. W., Liu, Z., Jones, J. F. & Zhang, J. (2018). Dietary intake of isoflavones and coumestrol and the risk of prostate cancer in the prostate, lung, colorectal and ovarian cancer screening trial. *International Journal of Cancer*, 142(4), 719–728.

Rodríguez-Landa, J. F., Cueto-Escobedo, J., Puga-olguín, A., Rivadeneyra-Domínguez, E., Bernal-Morales, B., Herrera-Huerta, E. V. & Santos-Torres, A. (2017). The phytoestrogen genistein produces similar effects as 17beta-estradiol on anxiety-like behavior in rats at 12 weeks after ovariectomy. *BioMed Research International*, 9073816. https://doi.org/10.1155/2017/9073816.

Ronis, M. J. (2016). Effects of soy containing diet and isoflavones on cytochrome P450 enzyme expression and activity. *Drug Metabolism Reviews*, 48(3), 331–341.

Rossi, M., Negri, E., Talamini, R., Bosetti, C., Parpinel, M., Gnagnarella, P., Franceschi, S., Dal Maso, L., Montella, M., Giacosa, A. & La Vecchia, C. (2006). Flavonoids and colorectal cancer in Italy. *Cancer Epidemiology, Biomarkers and Prevention*, 15(8), 1555–1558.

Russo, M., Spagnuolo, C., Tedesco, I. & Russo, G. L. (2010). Phytochemicals in cancer prevention and therapy: Truth or dare? *Toxins*, 2(4), 517–551.

Sak, K. (2017a). Current epidemiological knowledge about the role of flavonoids in prostate carcinogenesis. *Experimental Oncology*, 39(2), 98–105.

Sak, K. (2017b). Epidemiological evidences on dietary flavonoids and breast cancer risk: A narrative review. *Asian Pacific Journal of Cancer Prevention,* 18(9), 2309–2328.

Sakai, T. & Kogiso, M. (2008). Soy isoflavones and immunity. *Journal of Medical Investigation*, 55(3–4), 167–173.

Sakla, M. S., Shenouda, N. S., Ansell, P. J., Macdonald, R. S. & Lubahn, D. B. (2007). Genistein affects HER2 protein concentration, activation, and promoter regulation in BT-474 human breast cancer cells. *Endocrine*, 32(1), 69–78.

Satoh, H., Nishikawa, K., Suzuki, K., Asano, R., Virgona, N., Ichikawa, T., Hagiwara, K. & Yano, T. (2003). Genistein, a soy isoflavone, enhances necrotic-like cell death in a breast cancer cell treated with a chemotherapeutic agent. *Research Communications in Molecular Pathology and Pharmacology*, 113&114, 149–158.

Sehrawat, N., Yadav, M., Kumar, S. Upadhyay, S. K., Singh, M. & Sharma, A. K. (2020). Review on health promoting biological activities of mungbean: A potent functional food of medicinal importance. *Plant Archives*, 20(2), 2969–2975.

Setchell, K. D. R. & Lydeking, O. E. (2003). Dietary phytoestrogens and their effect on bone: Evidence from in vitro and in vivo, human observational, and dietary intervention studies. *American Journal of Clinical Nutrition*, 7, 593S–609S.

Setchell, K. D., Brown, N. M., Desai, P., Zimmer-Nechemias, L., Wolfe, B. E., Brashear, W. T., Kirschner, A. S., Cassidy, A. & Heubi, J. E. (2001). Bioavailability of pure isoflavones in healthy humans and analysis of commercial soy isoflavone supplements. *Journal of Nutrition*, 131(4), 1362S–1375S.

Setchell, K. D., Faughnan, M. S., Avades, T., Zimmer-Nechemias, L., Brown, N. M., Wolfe, B. E., Brashear, W. T., Desai, P., Oldfield, M. F., Botting, N. P. & Cassidy, A. (2003). Comparing the pharmacokinetics of daidzein and genistein with the use of 13C-labeled tracers in premenopausal women. *American Journal of Clinical Nutrition*, 77(2), 411–419.

Shen, F., Huang, W. L., Xing, B. P., Fang, X., Feng, M. & Jiang, C. M. (2018). Genistein improves the major depression through suppressing the expression of miR-221/222 by targeting connexin 43. *Psychiatry Investigation*, 15(10), 919–925.

Shi, H., Nam, P. K. & Ma, Y. (2010). Comprehensive profiling of isoflavones, phytosterols, tocopherols, minerals, crude protein, lipid, and sugar during soybean (Glycine max) germination. *Journal of Agricultural and Food Chemistry*, 58(8), 4970–4976.

Shiau, R. J., Chen, K. Y., Wen, Y. D., Chuang, C. H. & Yeh, S. L. (2010). Genistein and beta-carotene enhance the growth-inhibitory effect of trichostatin A in A549 cells. *European Journal of Nutrition*, 49(1), 19–25.

Shimazu, T., Inoue, M., Sasazuki, S., Iwasaki, M., Sawada, N., Yamaji, T., Tsugane, S. & Japan Public Health Center-based Prospective Study Group. (2010). Isoflavone intake and risk of lung cancer: A prospective cohort study in Japan. *American Journal of Clinical Nutrition*, 91(3), 722–728.

Shimazu, T., Inoue, M., Sasazuki, S., Iwasaki, M., Sawada, N., Yamaji, T., Tsugane, S. & JPHC Study Group. (2011). Plasma isoflavones and the risk of lung cancer in women: A nested case-control study in Japan. *Cancer Epidemiology, Biomarkers and Prevention*, 20(3), 419–427.

Shimoda, K., Kobayashi, T., Akagi, M., Hamada, H. & Hamada, H. (2008). Synthesis of oligosaccharides of genistein and quercetin as potential antiinflammatory agents. *Chemistry Letters*, 37(8), 876–877.

Si, H. & Liu, D. (2007). Phytochemical genistein in the regulation of vascular function: New insights. *Current Medicinal Chemistry*, 14(24), 2581–2589.

Sohn, S. I., Kim, Y. H., Kim, S. L., Lee, J. Y., Oh, Y. J., Chung, J. H. & Lee, K. R. (2014). Genistein production in rice seed via transformation with soybean IFSgenes. *Plant Science*, 217&218, 27–35.

Somjen, D., Amir-Zaltsman, Y., Gayer, B., Kulik, T., Knoll, E., Stern, N., Lu, L. J., Toldo, L. & Kohen, F. (2002). 6-carboxymethyl genistein: A novel selective oestrogen receptor modulator (SERM) with unique, differential effects on the vasculature, bone and uterus. *Journal of Endocrinology*, 173(3), 415–427.

Spagnuolo, C., Russo, G. L., Orhan, I. E., Habtemariam, S., Daglia, M., Sureda,A., Nabavi, S. F., Devi, K. P., Loizzo, M. R., Tundis, R. & Nabavi, S. M. (2015). Genistein and cancer: Current status, challenges, and future directions. *Advances in Nutrition*, 6(4), 408–419.

Squadrito, F., Marini, H., Bitto, A., Altavilla, D., Polito, F., Adamo, E. B., D'Anna, R., Arcoraci, V., Burnett, B. P., Minutoli, L., Di Benedetto, A., Di Vieste, G., Cucinotta, D., de Gregorio, C., Russo, S., Corrado, F., Saitta, A., Irace, C., Corrao, S. & Licata, G. (2013). Genistein in the metabolic syndrome: Results of a randomized clinical trial. *Journal of Clinical Endocrinology and Metabolism*, 98(8), 3366–3374.

Steensma, A., Faassen-Peters, M. A., Noteborn, H. P. & Rietjens, I. M. (2006). Bioavailability of genistein and its glycoside genistin as measured in the portal vein of freely moving unanesthetized rats. *Journal of Agricultural and Food Chemistry*, 54(21), 8006–8012.

Szeja, W., Grynkiewicz, G. & Rusin, A. (2017). Isoflavones, their glycosides and glycoconjugates. Synthesis biol. Activity. *Current Organic Chemistry*, 21(3), 218–235.

Szkudelska, K. & Nogowski, L. (2007). Genistein– A dietary compound inducing hormonal and metabolic changes. *Journal of Steroid Biochemistry and Molecular Biology*, 105(1–5), 37–45.

Tamura, M., Ohnishi-Kameyama, M., Nakagawa, H. & Tsushida, T. (2007). Dihydrogenisteinproducing bacterium TM-40 isolated from human feces. *Food Science and Technology Research*, 13(2), 129–132.

Tang, J., Xu, N., Ji, H., Liu, H., Wang, Z. & Wu, L. (2011). Eudragit nanoparticles containing genistein: Formulation, development, and bioavailability assessment. *International Journal of Nanomedicine*, 6, 2429–2435.

Teede, H. J. (2007). Sex hormones and the cardiovascular system: Effects on arterial function in women. *Clinical and Experimental Pharmacology and Physiology*, 34(7), 672–676.

Tham, D. M., Gardner, C. D. & Haskell, W. L. (1998). Clinical review 97: Potential health benefits of dietary phytoestrogens: A review of the clinical, epidemiological, and mechanistic evidence. *Journal of Clinical Endocrinology and Metabolism*, 83(7), 2223–2235.

Thangavel, P., Puga-olguín, A., Rodríguez-Landa, J. F. & Zepeda, R. C. (2019).Genistein as potentialtherapeutic candidate for menopausal symptoms and other related diseases. *Molecules*, 24(21), 3892.

Tian, T., Li, J., Li, B., Wang, Y., Li, M., Ma, D. & Wang, X. (2014). Genistein exhibits anti-cancer effects via down-regulating FoxM1 in H446 small-cell lung cancer cells. *Tumour Biology*, 35(5), 4137–4145.

Tosi, G., Costantino, L., Rivasi, F., Ruozi, B., Leo, E., Vergoni, A. V., Tacchi, R., Bertolini, A., Vandelli, M. A. & Forni, F. (2007). Targeting the central nervous system: In vivo experiments with peptide-derivatized nanoparticles loaded with loperamide and Rhodamine-123. *Journal of Controlled Release*, 122(1), 1–9.

Tuli, H. S., Tuorkey, M. J., Thakral, F., Sak, K., Kumar, M., Sharma, A. K., Sharma, U., Jain, A., Aggarwal, V. & Bishayee, A. (2019). Molecular mechanisms of action of genistein in cancer: Recentadvances.*Frontiers in Pharmacology*, 10, 1336. https://doi.org/10.3389/fphar.2019.01336.

Tuli, H.S., Mittal, S., Aggarwal, D., Parashar, G., Parashar, N. C., Upadhyay, S. K., Barwal, T.S., Jain, A., Kaur, G., Salva, R., Sak, K., Kumar, M., Varol., M., Iqubal, A. & Sharma, A.K. (2021). Path of silibinin from diet to medicine: A dietary polyphenolic flavonoid having potential anti-cancer therapeutic significance. *Seminars in Cancer Biology*, 73, 196–218.

Uifălean, A., Schneider, S., Ionescu, C., Lalk, M. & Iuga, C. A. (2015). Soy isoflavones and breast cancer cell lines: Molecular mechanisms and future perspectives. *Molecules*, 21(1), E13.

Van Erp-Baart, M. A., Brants, H. A., Kiely, M., Mulligan, A., Turrini, A., Sermoneta, C., Kilkkinen, A. & Valsta, L. M. (2003). Isoflavone intake in four different European countries: The Venus approach. *British Journal of Nutrition*, 89 (1), S25–S30.

Vauzour, D., Vafeiadou, K., Rice-Evans, C., Cadenas, E. & Spencer, J. P. E. (2007). Inhibition of cellular proliferation by the genistein metabolite 5, 7, 39, 49-tetrahydroxyisoflavone is mediated by DNA damage and activation of the ATR signalling pathway. *Archives of Biochemistry and Biophysics*, 468(2), 159–166.

Wang, S. D., Chen, B. C., Kao, S. T., Liu, C. J. & Yeh, C. C. (2014). Genistein inhibits tumor invasion by suppressing multiple signal transduction pathways in human hepatocellular carcinoma cells. *BMC Complementary and Alternative Medicine*, 14, 26. https://doi.org/10.1186/1472-6882-14-26.

Xiao, X., Liu, Z., Wang, R., Wang, J., Zhang, S., Cai, X., Wu, K., Bergan, R. C., Xu, L. & Fan, D. (2015). Genistein suppresses FLT4 and inhibits human colorectal cancer metastasis. *Oncotarget*, 6(5), 3225–3239.

Yadav, M., Sehrawat, N., Singh, M., Upadhyay, S. K., Aggarwal, D. & Sharma, A. K. (2020). Cardioprotective and hepatoprotective potential of citrus flavonoid naringin:

Current status and future perspectives for health benefits. *Asian Journal of Biological and Life Sciences,* 9(1), 1–5.

Yadav, M., Sehrawat, N., Upadhyay, S. K. & Kumar, S. (2021a). Emerging renoprotective role of citrus flavonoid naringin: Current pharmaceutical status and future perspectives. *Current Pharmacology Reports*, 7(3), 96–101.

Yadav, M., Upadhyay, S. K., Singh, R., Sharma, A. K., Kumar, S., Rani, K. & Sehrawat, N. (2021b). Effects of salinity stress on seedling growth of local varieties of faba bean (*Vicia faba* L.) from India. *Plant Archives*, 21(1), 253–259.

Yamamoto, S., Sobue, T., Kobayashi, M., Sasaki, S., Tsugane, S. & Japan Public Health Center-Based Prospective Study on Cancer Cardiovascular Diseases Group. (2003). Soy, isoflavones, and breast cancer risk in Japan. *Journal of the National Cancer Institute*, 95(12), 906–913.

Yan, G. R., Xiao, C. L., He, G. W., Yin, X. F., Chen, N. P., Cao, Y. & He, Q. Y. (2010). Global phosphoproteomic effects of natural tyrosine kinase inhibitor, genistein, on signaling pathways. *Proteomics*, 10(5), 976–986.

Yang, Y., Zang, A., Jia, Y., Shang, Y., Zhang, Z., Ge, K., Zhang, J., Fan, W. & Wang, B. (2016). Genistein inhibits A549 human lung cancer cell proliferation via miR-27a and MET signaling. *Oncology Letters*, 12(3), 2189–2193.

Yang, Z., Kulkarni, K., Zhu, W. & Hu, M. (2012). Bioavailability and pharmacokinetics of genistein: Mechanistic studies on its ADME. *Anti-Cancer Agents in Medicinal Chemistry*, *12*(10), 1264–1280.

Yeh, T. C., Chiang, P. C., Li, T. K., Hsu, J. L., Lin, C. J., Wang, S. W., Peng, C. Y. & Guh, J. H. (2007). Genistein induces apoptosis in human hepatocellular carcinomas via interaction of endoplasmic reticulum stress and mitochondrial insult. *Biochemical Pharmacology*, 73(6), 782–792.

Yoon, K., Kwack, S. J., Kim, H. S. & Lee, B. M. (2014). Estrogenic endocrine-disrupting chemicals: Molecular mechanisms of actions on putative human diseases. *Journal of Toxicology and Environmental Health. Part B, Critical Reviews*, 17(3), 127–174.

Yu, O., Jung, W., Shi, J., Croes, R. A., Fader, G. M., McGonigle, B. & Odell, J. T. (2000). Production of the isoflavones genistein and daidzein in non-legume dicot and monocot tissues. *Plant Physiology*, 124(2), 781–794.

Yu, Z., Li, W. & Liu, F. (2004). Inhibition of proliferation and induction of apoptosis by genistein in colon cancer HT-29 cells. *Cancer Letters*, 215(2), 159–166.

Yuan, J. P., Liu, Y. B., Peng, J., Wang, J. H. & Liu, X. (2009). Changes of isoflavone profile in the hypocotyls and cotyledons of soybeans during dry heating and germination. *Journal of Agricultural and Food Chemistry*, 57(19), 9002–9010.

Zengin, G. (2011). Synthesis, antimicrobial activity, and structure–activity relationships of eugenol, menthol, and genistein esters. *Chemistry of Natural Compounds*, 47(4), 550–555.

Zhang, H., Liu, G., Zeng, X., Wu, Y., Yang, C., Mei, L., Wang, Z. & Huang, L. (2015). Fabrication of genistein-loaded biodegradable TPGS-b-PCL nanoparticles for improved therapeutic effects in cervical cancer cells. *International Journal of Nanomedicine*, 10, 2461–2473.

Zhang, L. N., Xiao, Z. P., Ding, H., Ge, H. M., Xu, C., Zhu, H. L. & Tan, R. X. (2007). Synthesis and cytotoxic evaluation of novel 7-o-modified genistein derivatives. *Chemistry and Biodiversity*, 4(2), 248–255.

Zhang, Y., Hendrich, S. & Murphy, P. A. (2003). Glucuronides are the main isoflavone metabolites in women. *Journal of Nutrition*, 133(2), 399–404.

Zheng, X., Yao, X., Liu, Y., Zheng, Z., Cao, J. & Liao, D. (2010). Synthesis and cytotoxic activity of genistein derivatives. *Medicinal Chemistry Research*, 19(9), 1296–1306.

Zhou, B. P., Liao, Y., Xia, W., Spohn, B., Lee, M. H. & Hung, M. C. (2001a). Cytoplasmic localization of p21Cip1/WAF1 by Akt-induced phosphorylation in HER-2/neu-overexpressing cells. *Nature Cell Biology*, 3(3), 245–252.

Zhou, B. P., Liao, Y., Xia, W., Zou, Y., Spohn, B. & Hung, M. C. (2001b). HER-2/neu induces p53 ubiquitination via Akt-mediated MDM2 phosphorylation. *Nature Cell Biology*, 3(11), 973–982.

Zhu, Q., Meisinger, J., Van Thiel, D. H., Zhang, Y. & Mobarhan, S. (2002). Effects of soybean extract on morphology and survival of Caco-2, SW620, and HT-29 cells. *Nutrition and Cancer*, 42(1), 131–140.

Chapter 8

Delphinidin: Dietary Sources, Functional Properties and Health Benefits

Deeksha Sharma[1], Siddhartha Dan[2], Subashani[3], Kiran Upadhyay[4] and Sushil Kumar Upadhyay[5,*]

[1]Shoolini University of Biotechnology and Management Sciences, Shoolini University, Solan, Himachal Pradesh, India
[2]Department of Biotechnology, I.K. Gujral Punjab Technical University, Jalandhar, Punjab, India
[3]Department of Biochemistry, Chhatrapati Sahu Ji Maharaj University, Kanpur, Uttar Pradesh, India
[4]Department of Zoology, K.P. Higher Education Institute, Jhalwa-Prayagraj, Uttar Pradesh, India
[5]Department of Biotechnology, Maharishi Markandeshwar (Deemed to be University), Mullana-Ambala, Haryana, India

Abstract

There is escalating substantiation that polyphenols bring into being from natural products may impede with a number of processes concerned to many health challenges including cancer and metastasis. The epidemiological reports have exposed that a diet rich in fruits and vegetables has a beneficial anticipatory effect on cardiovascular diseases, neurodegenration, obesity and cancer by mechanisms that have not yet been elucidated very well. Delphinidin (Del or 3,3',4',5,5',7-hexahydroxyflavylium) and its glycosides are water-soluble pigments, belonging to anthocyanidins, a subgroup of flavonoids. Among 31 known anthocyanidin, the delphinidin abundantly identified in pigmented flowers, vegetables, beans, fruits particularly blueberry having a diphenylpropane-based polyphenolic ring structure and

* Corresponding Author's Email: upadhyay.k.sushil@gmail.com.

In: Flavonoids
Editors: Nirmala Sehrawat, Diwakar Aggarwal, Sushil Kumar Upadhyay et al.
ISBN: 978-1-68507-953-6

immense health promoting properties. These are health-promoting candidatures for pharmaceutical and nutraceutical applications, as substantiated by exhibiting antioxidation, anti-inflammation, antimicroorganism, antidiabetes, antiobesity, cardiovascular protection, neuroprotection, and anticancer potentials. The aforesaid actions of delphinidin are mediated through suppression of Met receptor phosphorylation, vascular endothelial growth factor receptor-2 phosphorylation, cancer cell proliferation and platelet-derived growth factor ligand/receptor signalling. However, additional investigation on the fundamental mechanisms of Del and its glycosides in mediating cellular processes with high specificity are still indispensable. In the present study, authors compiled the dietary sources, functional properties and health benefits of delphinidin (Dp) and a glycosylated form of delphinidin [delphinidin 3-O-beta-glucopyranoside (Dp 3-glu)].

Keywords: polyphenols, flavonoids, delphinidin, pharmaceutical, nutraceutical, anticancer

1. Introduction

In plants, anthocyanins are included in flavonoid class, which considered as largest group of water-soluble, visible pigments (Khoo et al., 2017; Sehrawat et al., 2020; Yadav et al., 2020; Aggarwal et al., 2021; Tuli et al., 2021; Yadav et al., 2021a,b). From 31 anthocyanidins, more than 500 anthocyanins have been identified and isolated so far (McCallum et al., 2007). Among these, only 6 anthocyanidins that include cyanidin, peonidin, delphinidin, pelargonidin, petunidin and malvidin, are pervasive and having dietary importance (Tanaka et al., 2008). The six anthocyanidins having same cellular actions and similar metabolism, but there is only one structural difference among these anthocyanidins noticed as the addition of hydrogen, methyoxyl or hydroxyl group at C-3′ and C-5′ positions. Various evidences indicated that among nutraceuticals and food supplements, the popularity of anthocyanins continuously increasing due to beneficial their effects. According to an ad-hoc food composition database, Phenol-Explorer and US Department of Agriculture databases, the estimation of mean intake of total amount of anthocyanidin in men is ranged between 19.83-84.88 mg/d however among women about18.73–44.08 mg/d (Zamora-Ros et al., 2011). It is shown in increasing studies that health promoting anthocyans (anthocyanidins and anthocyanins) considered to have the potential to protect from various human diseases (Patel et al., 2013; Khoo et al., 2017; Naseri et al., 2018).

Anthocyans showed a wide variety of pharmacological effects which includes anti-inflammation (Lee et al.,2014), anti-diabetes (Mojica et al., 2017; Naseri et al., 2018), antioxidant (Wang et al., 2016), antimicrobial (Jimenez et al., 2010; Santos et al., 2011; Hariri et al., 2016), cardiovascular (Jana et al., 2017), prevention of muscle atrophy (Murata et al., 2016), anticancer (Bishayee et al., 2010) and neural protection (Kim et al., 2012a,b). Delphinidin (Del) is a major anthocyanidins present in berries, flowers, fruits and beans, having a diphenylpropane-based polyphenolic ring structure and well documented for having health promoting properties. Stability, transformation and bioavailability are contributed due to secondary modifications in the structure of delphinidin. For example, *in vivo*, increase in the stability is due to glycosylation modification of Delphinidin, influenced by biochemical and physicochemical factors. Few natural acylated Del glycosides are identified and isolated (Sakaguchi et al., 2013). Though, in naturally occurring small compounds, glycosylation modifications are more often seen. In this article, few artificial and synthetic Del derivatives (Legeay et al., 2019) that exhibit potent biological activities are also included. Zhao et al. (2017) examined the chemical structure-based implications in the stability of anthocyanins, including Del glycosides. The glycosyl substitution can be produced as a variety of Del glycosides, including delphinidin-3-arabinoside (Del-3-ara), delphinidin-3-glucoside (Del-3-glu), delphinidin-3-rutinoside (Del-3-rut), delphinidin-3-galactoside (Del-3-gal), delphinidin-3-sambubioside (Del-3-sambubioside) (Del-3-sam) and delphinidin-3-rutinoside (Del-3-rut).

Glucuronide conjugation significantly extends the peak timeperiod of Del, increasing its bioavailability (Zhong et al., 2017). Del and its glycosides have a variety of biological actions that are strongly related to their chemical structures. Del and its glycosides have a wide range of biological actions that are strongly related to their chemical structures. The pyrogallol moiety in Del confers the ability to decrease the expression of the IgE receptor (FcRI) in human mast cells, relieving allergy symptoms, according to SAR research (Tamura et al., 2010). In this article, the nutraceutical and pharmaceutical uses that were reported from last 10 years has been mainly discussed with respect to Del and its glycosides. Del and its glycosides belong to subclass of flavonoids and are water soluble. The health prompting effect of Del and its glycosides are best candidates in nutraceutical and pharmaceutical industries because of their anti-inflammation, anti-microbial, anti-oxidant, anti-obesity, anti-cancerous, cardiovascular protection, anti-diabetes and neuroprotection. Del glycosylation modification is linked to decreased biological activity and increased stability. Del and its glycosides have the potential to be alternative inhibitors of ER, CBRs, BCRP, SGLT-1 and EGFR and virtual docking suggests that the sugar moiety may not interact well with the targets' active sites.

Del and its glycosides have multifunctional capabilities due to structural features. Del and its glycosides are promising and have been researched as prospective medicines due to their health-promoting properties. More research on the underlying mechanisms of Del and its glycosides in influencing cellular activities with great selectivity is required (Chen et al., 2019a, b). Anthocyanins are soluble pigments in water that are responsible for the red–blue colour of many fruits and vegetables including grapes and their derivatives, such as wines. As the fruit ripens, the increase in the concentration of these anthocyanins have been recorded (Calderón-Oliver and Ponce-Alquicira et al., 2018).In MCF-10F cells, a noncancerous, immortalised human breast epithelial cell line, anthocyanin-rich extracts from Concord grapes and delphinidin (anthocyanin) both generated a large increase in GST levels (Singletary et al., 2007). On contrary, fractions containing anthocyanins from Georgia-grown blueberries, with major components delphinidin, cyanidin, peonidin, petunidin, and malvidin, all at concentrations (50–150 g/mL), were shown to lower GST activity in HT-29 cells, with a concurrent decrease in quinone reductase activity (Srivastava et al., 2007). In various positions of the flavonoid ring are found to be linked to sugars such as glucose, rhamnose, xylose, arabinose, and fructose.

Anthocyanidins are compounds that have an aglycone or a structure that does not contain sugar. Delphinidin, malvidin, pelargonin, peonidin, Cyanidin, pelargonin, petunidin are some of the most prevalent anthocyanidins. Cyanidin is the primary pigment found in fruits including plums, cherries, and berries. The primary pigment found in strawberries and radish is perlagonin. Delphinidin is the primary anthocyanidin found in eggplant as well. Malvidin is the most common anthocyanidin found in red grapes and wine. It has been observed that cyanidin-3-rutinoside, inhibits part of the damage induced by high blood glucose levels, such as protein glycosylation and oxidative damage to proteins and DNA, by trapping free radicals (Thilavech et al., 2016). *In vitro* studies showed that cyanidin glucosides reduce oxidative stress and hence pancreatic beta cell deaths were documented (Lee et al., 2015). Malvidin, like other polyphenols, promotes anti-inflammatory pathways (Huang et al., 2014).

Cyanidin glucosides have been shown to reduce the growth of prostate cancer cell lines, reduce the levels of ROS and enhance tumour suppressive proteins such as P75NGFR (Sorrenti et al., 2015).In diabetic rats, a bayberry extract high in cyanidin-3-glucosides reduces hyperglycemia and pancreatic - cell damage (Sun et al., 2012a,b). Delphinidin glucosides isolated from black currant decrease cancer cell growth (*in vitro*) *via* processes dependent on apoptosis and alterations in signalling pathways such as Akt/Bad/Bcl-2 (León González et al., 2015; Kopustinskiene et al., 2020). It appears that fruits or fruit juice extracts contain high levels and different kinds of anthocyanins, have a significant impact on health (Sehrawat et al., 2020; Yadav et al., 2020; Aggarwal et al., 2021; Tuli et al., 2021; Yadav et al., 2021a,b). *In vitro* studies have shown that extracts from berries containing anthocyanins, such as cyaniding and delphinidin glucosides, enhance insulin secretion and alter the expression of genes and proteins related to insulin. Furthermore, anthocyanins isolated from Java plum reduced cancer cell growth (Charepalli et al., 2016; Johnson and De Mejia, 2016). Black currant and blueberry extracts, which include malvidin, petunidin, and delphinidin glucosides, are another example; depending on the dosage, they inhibit the growth of certain cancer cells(Kausar et al., 2012; Diaconeasa et al., 2015).

Anthocyanins may lower the chance of developing cancer or the secondary consequences of the disease, such as metastasis (Kausar et al., 2012; Diaconeasa et al., 2015). Blueberry extracts promote apoptosis and the activity of enzymes such as lactate dehydrogenase in melanoma cells, decreasing their growth (Bunea et al., 2013). Despite the combination of different polyphenols, in this instance anthocyanins, and medications, such as used in chemo-therapies, it provides a synergic impact that reduces cancer cell growth (Singh et al., 2016; Lin et al., 2017).Many anthocyanins are metabolised by the gut microbiota, which create metabolites with health-promoting properties, such as hydroxycinnamic, hydroxybenzoic, and other acids (Aura et al., 2005; Keppler and Humpf, 2005; Fang, 2015). The anthocyanin consumption may be helpful to a lower risk of cardiovascular disease as well (Wallace, 2011).

2. Dietary Source of Delphinidin

Delphinidin (2-(3, 4, 5-trihydroxyphenyl) chromenylium-3, 5, 7-triol) is a diphenylpropane-based polyphenolic ring structure found in the vacuolar sap of flowers and the epidermal tissues of fruits (Table 1). Delphinidin was shown to have antioxidant, antimutagenic, anti-inflammatory, and antiangiogenic effects (Satué-Gracia et al., 1997; Wang et al., 1997). The above-mentioned actions of delphinidin are mediated through suppression of Met receptor phosphorylation, vascular endothelial growth factor receptor-2 phosphorylation, cancer cell proliferation and platelet-derived growth factor ligand/receptor signalling (Domitrović et al., 2010; Feng et al., 2010; Kumoro et al., 2010; Takasawa et al., 2010). Delphinidin may be found in a variety of brightly coloured vegetables and fruits, as well as nutritional supplements that are now being used as alternative cancer treatment (Ozbay and Nahta, 2011; Patel et al., 2013).

Table 1. Dietary source of delphinidin and its concentration in different fruits

Compounds	Fruits	Contents	Reference(s)
Delphinidin	Black currant (mean of different varieties)	3049.1 ± 787.0 mg/kg FW (as delphinidin 3-O-rutinoside)	Grumezescu and Holban 2017
	Banana (*Musa acuminata* Colla)	73.9 ± 11.8 mg/kg FW	Bhagwat et al., 2013
	Bilberry or European blueberry (*Vaccinium myrtillus*)	975.9 ± 50.5 mg/Kg FW	Bhagwat et al., 2013

3. Pharmacological and Health Benefits Potential of Delphinidin

Delphinidin is an anthocyanidin found in abundance in coloured vegetables and fruits, notably blueberries, and known for its antioxidant and anti-inflammatory effects. The effects of delphinidin on psoriatic epidermal keratinocyte inflammation, proliferation and differentiation were recently studied utilising a rebuilt human psoriatic skin equivalent (PSE) model (Figure 1). Psoriasis lesions lose their histological characteristics, and neutrophil and macrophage infiltration is greatly reduced.

Psoriasis is a prevalent skin condition that causes both physical and psychological distress. Although there is evidence of a hereditary susceptibility, the cause is unknown. The involvement of the immune system in the development of psoriasis is also a prominent area of study. Although it has been suggested that psoriasis is an autoimmune disease, no autoantigen responsible for the illness has been identified. Psoriasis can be triggered by both external and internal factors, including as stress, minor trauma, infections, systemic medications and sunburn (Bonesi et al., 2018).

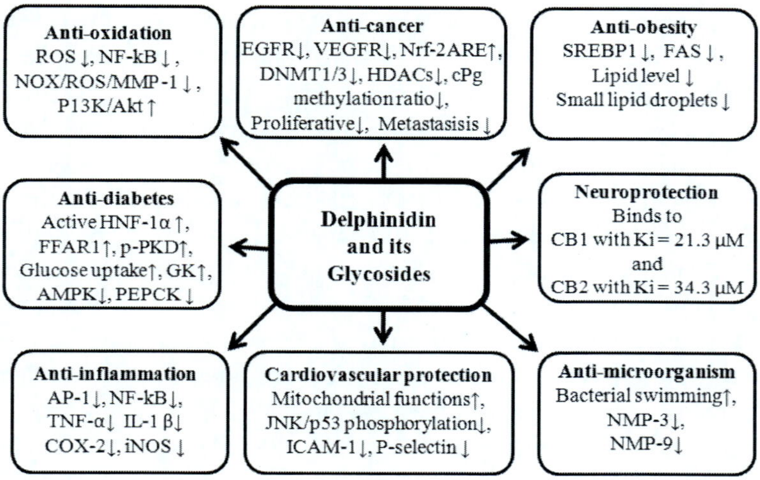

Figure 1. Pharmacological and health benefits potential of delphinidin.

Psoriasis is found all around the world. It affects both sexes of all ages in all nations, regardless of ethnic heritage. The prevalence of psoriasis in different nations ranges from 0.09 percent to 11.4% in the published data. In most developed nations, the prevalence is between 1.5 and 5%. There is also evidence that suggested that the prevalence of psoriasis is on the rise.

One recent study found that men have more severe manifestations of the disease than in women. On 24th May, 2014, the World Health Organization (WHO) passed a resolution on psoriasis at its 67thWorld Health Assembly (WHA 67.9). The World Health Organization acknowledges the urgent need to support international initiatives to promote awareness about the disease and combat stigmatization of people suffering from the psoriasis. It can manifest in many different forms. The addition of delphinidin to PSE enhanced keratinocyte differentiation. Cornification induced by delphinidin (without altering apoptosis) as well as the expression of protein and mRNA markers of differentiation markers such as filaggrin, loricrin, involucrin and caspase-14. It also reduced the expression of inflammatory markers often seen in psoriatic skin, such as the antimicrobial peptides S100A7-psoriasin and S100A15-koebnerisin. Pal et al. (2015) investigated the potential of topically administered delphinidin to alter pathological indicators of psoriasis lesions in a flaky skin mice model.

3.1. Anti-Cancer Activity of Delphinidin

The impact of delphinidin on established breast cancer cell lines of distinct molecular subtypes was studied to evaluate anticancer efficacy. Delphinidin reduced proliferation, hindered anchorage-independent growth, and caused apoptosis in ER-positive, triple-negative, and HER2-overexpressing breast cancer cell lines. After treatment with delphinidin, MAPK (mitogen-activated protein kinase) signalling was decreased in triple negative cells and ER-negative chemically altered MCF10A cells (Patel et al., 2013). Anthocyanins have received a lot of interest due to their wide variety of possible health benefits. The mechanisms of action of delphinidin's and other anthocyanins' cytoprotective effects were explored. Through the suppression of many critical signalling cascades, all of the tested flavonoids were shown to prevent peroxynitrite-induced apoptotic effects on endothelial cells. Furthermore, epidemiological investigations have shown that delphinidin is beneficial at various stages of carcinogenesis (Clere et al., 2011; Paixão et al., 2011).

Another research found that the expression of tyrosyl-DNA phosphodiesterase 1 had no effect on the degree of DNA damage caused by delphinidin, showing that the induction of DNA strand breaks is not primarily topoisomerase-mediated. The addition of catalase to the cell culture medium, however, reduced the DNA-damaging effects of delphinidin by counteracting delphinidin-mediated hydrogen peroxide production (Fritz et al., 2008). It was also reported that delphinidin suppresses tumour promoter-induced transformation and cyclooxygenase-2 (Cox-2) production in JB6 P+ mouse skin epidermal cells by directly mitogen-activated protein kinase (MEK) and targeting Raf (Figure 1). TPA activation of activator protein-1 and nuclear factor-kappaB was shown to be dosage dependent, and it is also blocked by delphinidin (Kang et al., 2008).

The effects of delphinidin on human colon cancer HCT116 cells were studied (Table 2). Delphinidin induced apoptosis, activated caspases-3, -8, and -9,increase in Bax, cleaved PARP, reduced cell viabilityin Bcl-2 protein, and arrested the cell cycle in the G2/M phase (Yun et al., 2009). Another study investigated at the effects of anthocyanidins on primary (Caco-2) and metastatic (LoVo and LoVo/ADR) colorectal cancer cell lines. Delphinidin was shown to be cytotoxic to metastatic cells (Cvorovic et al., 2010). The effects of anthocyanins on various mechanisms of cell death in various malignancies were studied. Delphinidin repressed UVB-induced Cox-2 expression in JB6 P+ mouse epidermal cells. Cox-2 promoter activity and PGE (2) formation were further reduced by delphinidin treatment within noncytotoxic amounts (Kwon et al., 2009). The effect of delphinidin therapy on human C4-2, PC3 cells, 22Rnu1 and PCa LNCaP was examined, and it was discovered that it inhibited cell proliferation in a dose-dependent manner without harming normal human prostate epithelial cells (Hafeez et al., 2008). Delphinidin promotes apoptosis and cell cycle arrest in androgen-refractory human PCa 22Rnu1 cells, and the effects are concurrent with NfkappaB inhibition (Bin Hafeez et al., 2008).

Table 2. Functional health benefits of delphinidin with reference to its concentration

Compounds	Concentrations	Health Benefits	Reference(s)
Delphinidin	25–200 µg/mL in human umbilical vein endothelial cells	Decrease ROS, reduce theoxidation of low-density lipoproteins, could prevent plaque development in atherosclerosis disease	Yao, et al., 2017
	30–240 µM in human colon cancer HCT116 cells 100 g/kg in diabetic mice at eight weeks	Induce apoptosis activating caspases 3, 8, 9; cause cycle arrest and decrease albumin glycation	Yun et al., 2009; Gharib et al., 2013

3.2. Anti-Inflammatory Activity of Delphinidin

Delphinidin (derived from *Punica granatum* L) hindered the Histone acetyltransferase (HAT) activities of p300/CBP particularly. It also reduced p65 acetylation in MH7A cells, a synovial cell line from patients with rheumatoid arthritis. TNF-stimulated elevations in NF-B function and expression of NF-kB target genes in these cells were reduced by delphinidin administration. Delphinidin inhibited the production of pro-inflammatory cytokines produced by lipopolysaccharide in Jurkat T cells, indicating that HATi effectively inhibits cytokine-mediated immune responses (Seong et al., 2011). Oxidative and inflammatory stress may have a beneficial interaction (Lugrin et al., 2014). Anthocyanins have several biological functions, including anti-inflammation. Cocoplum anthocyanin extracts, including Del-3-glu, Del-3-ara, and Del-3-gal, have been shown to elevate TNF-, IL-1, IL-6, and NF-B expression and decrease intracellular ROS generation in HT-29 cells while having no cytotoxic effects on non-malignant colonic fibroblasts CCD18Co cells (Venancio et al., 2017).

Furthermore, Del and Del-3-sam suppress the expression of iNOS, NO, IL-6, MCP-1, and TNF-, as well as the activity of the LPS-induced NF-B and MEK1/2-ERK1/2 signalling pathways (Table 1) (Sogo et al., 2015). Del-3-rut inhibits the expression of IL-1 mRNA, the translocation of NF-B, and the generation of ROS in bone marrow derived macrophages (BMMs) from wild-type mice. The effects of Del-3-rut on NF-B/IL-1 signalling, however, are not impaired in Nrf2-deficient BMMs. This implies, Del-3-rut suppresses LPS-induced NF-B translocation without requiring NRF2 activation (Lee et al., 2014). Nasal polyps are distinguished by mucous membrane irritation and extracellular matrix buildup (ECM). Del has been shown to reduce the production of -SMA, fibronectin, and collagen in nasal polyp-derived fibroblasts by inactivating the MAPK/NF-B signalling pathway (Cho et al., 2015).

Pro-inflammatory substances secreted, such as cytokines and chemokines, play critical roles in the pathogenesis of chronic obstructive pulmonary disease (COPD). Cigarette smoke extract increases the expression of the neutrophil chemoattractant IL-8, complicating the development of COPD. Del and cyanidin (Cya) (1 M and 10 M) have been shown to suppress IL-8 production in a dose-dependent manner, reversing the effects of smoke exposure on human small airway epithelial (SAE) cells (Flores et al., 2012). The production of IL-4-induced eotaxin-3 (CCL26) and the migration of eosinophils into the airway have been substantially reduced by polyphenolic extracts of black currant cultivars, which are mainly composed of anthocyanins such as Del-3-glu and Del-3-rut. This inhibitory action is enhanced by a greater Del/Cya glycoside ratio (Nyanhanda et al., 2014).

Del substantially reduces LPS-induced MUC8 and MUC5B expression *via* inhibiting the TLR-4-regulated ERK1/2 and p38 MAPK signalling pathways in NCI-H292 human airway epithelial cells (Bae et al., 2014). Del has been reported to suppress the IL-1-induced NF-B/Cox-2 signalling pathway and prevent phosphorylation of IRAK-1 and NF-B-inducing kinase (NIK) in chronic osteoarthritic chondrocytes, leading in protection from inflammatory stress in human cartilage chondrocytes (Haseeb et al., 2013). Histone acetyltransferase (HAT) activation increases inflammation, which contributes to the development of rheumatoid arthritis. Del inhibits HAT activity, resulting in p65 hypoacetylation and cytosolic accumulation, suppression of the NF-B signalling pathway, and suppression of cytokine-mediated immunological responses in MH7A cells (Seong et al., 2011). Another research found that Del inhibits RANKL-induced osteoclastogenesis in RAW264 cells, as evidenced by reduced expression of cfos, Nfatc1, MMP9, and Trap and attenuated regulation of NF-B signalling (Moriwaki et al., 2014). Del-3-rut, on the other hand, has no impact on the expression of RANKL produced by t-BHP but has a significant reduction of ROS production and increase of GSH/GSSH, avoiding osteoblast dysfunction in age-related osteoporosis (Casati et al., 219).

Chronic gut inflammation can increase the permeability of the intestinal barrier. Anthocyanins, particularly Del-3-glu, have been demonstrated to protect Caco-2 cell monolayers from TNF-induced decreases in transepithelial electrical resistance (TEER) and increases in FITC-dextran permeability. The potential mechanism might be related with Del-3-glu reduction of TNF-induced NF-B signalling and MLC phosphorylation, resulting in the inhibition of tight junction factor downregulation and redistribution (Cremonini et al., 2017). Glucose (5.62 mM) may stimulate cell proliferation and collagen I and IV mRNA expression, as well as increase ROS generation, Nox-1 expression, and ERK1/2 and TGF- activity. Del has been shown to reduce the effects of glucose on mesangial cells (Table 1) (Song et al., 2016). Cox-1 is essential for thromboxane production and the preservation of gastrointestinal epithelial integrity. Five anthocyanidins, including cyanidin, malvidin, peonidin, pelargonidin, and Del, have been found to suppress Cox-1 and Cox-2 activity (Ricciotti and FitzGerald, 2011).

According to SAR investigations, the quantity and location of hydroxyl groups in ring B are critical for such inhibitory action. The function of C-3′ and C-4′ hydroxyl substitution in Cox inhibitory action appears to be essential. The hydroxylation of C-3′, C-4′, and C-5′ in Del, on the other hand, result in a decreased activity (Seeram et al., 2003). In vitro studies have demonstrated that prodelphinidins inhibit Cox-2/PGE2 activity with great selectivity. In the whole blood test, however, prodelphinidins had no influence on Cox activity (Garbacki et al., 2002). Del therapy substantially reduces the expression of Cox-2/PGE2 and the activities of p38MAPK/NF-B in spinal cord injury (SCI)-induced inflammation (Wang et al., 2017). UVB can stimulate Cox-2/PGE2 expression in JB6 P+ murine epidermal cells. Del may bind to and inhibit MAPKK4 or PI3K in a competitive way, resulting in the inactivation of UVB-induced JNK, p38, and Akt, as well as the abrogation of NF-κB/Cox-2 signalling (Kwon et al., 2009). Targeting the Cox-2 signalling system, which is regulated by TNF, could be an approach for controlling skin inflammation and cancer. Del also binds to Fyn kinase in a non-competitive manner and suppresses its activity in JB6 P+ cells, where it is engaged in the TNF alfa-mediated Cox-2 signalling pathway (Hwang et al., 2009). Surprisingly, some study suggests that Del may have pro-inflammatory properties. Del has been reported to promote the expression of IL-2 and IFN- *via* upregulating nuclear factor of activated T cells (NFAT) and Ca2+ release-activated Ca^{2+} (CRAC) channel activity in Jurkat cells and peripheral blood lymphocytes (Jara et al., 2014).

3.3. Antioxidant Activity of Delphinidin

In human umbilical vein endothelial cells (HUVECs), the effects of delphinidin on oxidised low-density lipoprotein (oxLDL)-induced damage were investigated. The viability loss produced by oxLDL in HUVECs was demonstrated to be considerably repaired by delphinidin in a concentration-dependent manner (Estévez et al., 2008; Chen et al., 2010). In the tested cell line, Delphindin increased the levels of p27, which were reduced by hydrogen peroxide. Hydrogen peroxide and delphinidin appear to modulate intracellular levels of p27 *via* modulating HIF-1 levels, which are governed by its upstream regulators, which include the PI3K/Akt/mTOR signalling pathway (Quintos et al., 2010). In another investigation, anthocyanidins were found to have substantial scavenging action against various free radicals, with delphinidin being the most active at the measured dose (Ogawa et al., 2011). The most prevalent radical species are hydroxyl and nitric oxide (NO). The former is thought to be the most reactive and harmful. Del may decrease hydroxyl radicals but hardly reacts with nitric oxide radicals (Dimitric et al., 2017).

Del has antioxidative action that protects APRE-19 cells from apoptosis produced by H_2O_2, as evidenced by reduced ROS generation, down-regulation of Nox1, and up-regulation of SOD, CAT, GSH-Px, and NRF2 (Ni et al., 2019). Del's antioxidative effect is linked to its own structures, specifically the hydroxyl components in the B-ring (C-4′). The oxygen radical absorbance capacity (ORAC) test is used to assess antioxidants' radical-scavenging action. Del-3-glu has an antioxidative capacity of 8.33 mol TE/mol, whereas Del-3-gal has a capacity of 13.062 mol TE/mol (Chorfa et al., 2016). This discrepancy suggests that the antioxidative efficacy of Del glycosides may be due to the sugar moieties' liability. Furthermore, glycosylation on Del C-3 has higher antioxidative effects to decrease ROS generation than diglycosylation on both C-3 and C5. The underlying mechanism of Del-3-glycosides in antioxidation may be strongly related to enhanced glutathione reductase protein expression levels in HT-29 cells (Jing et al., 2015).

The antioxidative efficacy of Del may change and dependson its concentration. Del has antioxidative action and protects cells from oxidative stress-induced apoptosis at low concentrations (100 nM to 1μM). Del has cytotoxic effects on human umbilical vein endothelial cells (HUVECs) at higher concentrations (100 M) (Goszcz et al., 2017). Another study found that Del, at a concentration of 100 g/mL, is cytotoxic to tenofibroblasts. Del, at lower concentrations, has been shown to decrease the formation of intracellular ROS and increase the activation of ERK1/2 and JNK, therefore reducing oxidative stress-induced tenofibroblast cell death (Nam et al., 2016). This contradiction might be explained by Del's intrinsic instability. It is unclear whether this antioxidative activity is a direct result of Del or an indirect result of its degradation. Anthocyanins have the potential to alter mitochondrial respiration and decrease oxidative stress. Del and its glycosides have been shown to sometimes uncouple oxidative phosphorylation, lower mitochondrial membrane potential, and diminish ROS generation in the mitochondria (Liobikas et al., 2016).

The Del protects embryonic stem (ES) cells by reducing ROS generation and phosphorylation of JNK and NF-B, as well as correcting hypoxia-induced decreases in phosphorylation of AktThr308 and Ser473. Furthermore, Del inhibits the production of Bax, caspase-9, and caspase-3 caused by hypoxia while increasing the expression of Bcl-2, therefore inhibiting cell death (Seo et al., 2013). Del-3-glu and Del-3-rut from black currant increase the phosphorylation of endothelial nitric oxide synthase (eNOS) and Akt in HUVECs in vitro (Edirisinghe et al., 2011). Hepatic steatosis has been related to oxidative stress caused by the buildup of free fatty acids in liver cells. Del-3-glu has hepatoprotective efficacy in HepG2 cells by attenuating oleic acid-induced hepatic steatosis, as shown by diminished triglyceride accumulation and triglyceride elevated clearance, as well as lower ROS generation (Calderón-Oliver and Ponce-Alquicira, 2018). TGF-1 can cause ROS production in liver fibrosis and cell death in implanted 2m-/Thy1+BDHSCs. Del activates the PI3K/Akt signalling pathway and modulates the expression of Bcl-2, Bax, and caspase-3 in 2m-/Thy1+BDHSCs, suppressing TGF-1-induced ROS production and inhibiting apoptosis (Chen et al., 2019a,b). Epigallocatechin-3-gallate (EGCG), a potent antioxidant, was shown to have a pro-oxidative activity. Del-3-gal may protect cells from EGCG-induced cytotoxicity and endoplasmic reticulum stress in mouse hepatocytes by increasing the expression of heme oxygenase-1 (HO-1) and heat shock protein-70 (HSP-70) (Inoue et al., 2012). UVB radiation (280320 nm) causes a variety of skin disorders. MMP-1 activation is related with cutaneous collagen degradation and collagen matrix disintegration. Del may block the phosphorylation of MKK4-JNK1/2, MKK3/6-p38, and MEK-ERK1/2, as well as the activity of NADPH oxidase (NOX), resulting in reduced synthesis of UVB-stimulated ROS and MMP-1 expression (Lim et al., 2013).

Similarly, Del has cytoprotective activity against radiation in HEL-299 human lung cells by enhancing the expression of antioxidant enzymes and pro-survival proteins while reducing the expression of pro-apoptosis factors (Kim et al., 2018). However, during analyzation for antioxidative activity against UVB-induced damage in ARPE-19 cells, Del-3-glu (5 M) did not show any antioxidative effects, as shown by no changes in MAPK phosphorylation (Silvan et al., 2016). Del exhibits cytotoxicity in drug-resistant LoVo/ADR colorectal cancer (CRC) cells via increasing ROS production and decreasing glutathione reductase (Cvorovic et al., 2010). Del may have pro-oxidative or anti-oxidative action depending on its concentration, cell type, context specificity of cellular responses, and the complex microenvironment in cells or tissues.

3.4. Hepatoprotective Activity of Delphinidin

The hepatoprotective effects of delphinidin in mice with carbon tetrachloride (CCl (4))-induced liver fibrosis were studied. Delphinidin has successfully reduced oxidative stress, elevated the expression of matrix metalloproteinase-9 and metallothione in I/II, and repaired hepatic architecture. The therapeutic action of delphinidin is primarily due to the deactivation of hepatic stellate cells (HSC) and the down-regulation of fibrogenic stimuli, resulting in a significant increase in hepatic regeneration potential (Domitrović et al., 2010).

3.5. Neuroprotection Activity of Delphinidin

Oxidative stress has been connected to the development of Alzheimer's disease (AD) and Parkinson's disease (PD). Del-3-glu reduces H2O2 cytotoxicity in SK-N-SH cells. The possible mechanisms for Del-3-glu in neuroprotection might be related with the inhibition of the ASK-JNK/p38 signalling pathway, the upregulation of HO-1, and enhanced mobilization of cellular free sialic acid (Kim et al., 2012a,b). Del consistently demonstrates neuroprotective properties against hypoxia, which may be connected to increased CAT and SOD activity as well as increased GSH concentration in C6 glial cells (Lin et al., 2014). The -amyloid peptide (A) builds up and forms a senile plaque, which is one of the hallmarks of Alzheimer's disease. A2535 reduces cell viability while increasing calcium levels and tau phosphorylation in PC12 cells. Del has been shown to impair cytotoxicity-induced A2535 by inhibiting intracellular calcium influx and tau phosphorylation (Kim et al., 2009). The fusion of released vesicles with the presynaptic plasma membrane is required for neurotransmitter transmission. The soluble N-ethylmaleimide-sensitive factor attachment protein receptor (SNARE) is responsible for this activity. In a study, it was found that certain tiny hydrophobic compounds have been identified to control SNARE-mediated membrane fusion via various action mechanisms. Del may impede the N-terminal nucleation of SNARE zippering, whereas myricetin binds to the hydrophobic inner core at the C-terminal, preventing SNARE zippering motion and trans-complex accumulation in PC12 cells.

These findings suggest that SNARE could be considered a specific target for treating diverse hypersecretion disorders (Yang et al., 2010; 2015). Anthocyans have been proven to have neuroprotective and analgesic effects. Del binds to human CB1 with a Ki value of 21.3 M and CB2 with a Ki value of 34.3 M (Figure 1), according to competitive radioligand binding experiments (Korte et al., 2009).

Delphinidin's neuroprotective properties against a beta-induced toxicity were studied. Delphinidin protected PC12 cells against a beta by lowering tau phosphorylation and intracellular calcium levels (Kim et al., 2009). The *in vitro* and *in vivo* study of effects of delphinidin isolated from *Vaccinium myrtillus* on retinal ganglion cells (RGCs) against retinal injury were studied. Delphinidin reduced radical activation in RGC-5 andSIN-1-induced neurotoxicity, as well as lipid peroxidation in mouse forebrain homogenates, in a concentration-dependent manner (Matsunaga et al., 2009). Neuronal SNARE proteins facilitate vesicle fusion with the presynaptic plasma membrane, therefore mediating neurotransmitter release at the synapse. Wedging with micro hydrophobic molecules can modify SNARE zippering in the middle, but delphinidin and cyanidin block SNARE zippering at the N-terminus (Yang et al., 2010).

3.6. Cardiovascular Protection Activity of Delphinidin

The abnormal development of cardiovascular illnesses is influenced by inflammatory and oxidative stress (Yang et al., 2019). Several epidemiological studies suggest that taking anthocyanidins reduces the overall risk of death and cardiovascular disease. This impact could be linked to anthocyanidins' ability to protect against inflammation and oxidative damage. Dayoub, et al. (2014) evaluated the positive effects of Del on cardiovascular endothelial cells. Endothelial cells are activated and driven to express cell adhesion molecules and recruit monocytes in response to inflammation, which is an important early stage in the formation of atherosclerosis. Del-3-glu may inhibit the activity of MEK2 and FAK1 by physical contact, according to molecular virtual docking (Krga et al., 2016; 2018a,b). Anthocyanins have been shown to protect cardiac cells from oxidative stress-induced apoptosis (Isaak et al., 2017).

Endothelial dysfunction has been shown to be inhibited by four flavonoid subclasses. In EA.hy926 human vascular endothelial cells, Del was found to be the most effective in reducing ROS production induced by oxidised low-density lipoprotein (ox-LDL), as well as inactivation of the p38 MAPK and NF-B signalling pathways and downregulation of ICAM-1, VCAM-1, E-selectin, and MMP-1/-2/-9 (Yi et al., 2012). Del-3-glu consistently counteracts the effects of ox-LDL on porcine aortic epithelial cells (PAEC), as evidenced by decreased ROS production, improved mitochondrial function, and decreased Nox2/4/p22phox expression (Xie et al., 2012). Del-3-glu also restores ox-LDL-induced damage in HUVECs by activating the AMPK/SIRT1 signalling pathway and, as a result, increasing autophagy activity (Jin et al., 2014). Del also suppresses the ROS/p38MAPK/NF-B signalling, lowering the expression of ox-LDL-induced adhesion factors (ICAM-1 and P-selectin) and avoiding atherosclerosis (Chen et al., 2011). Del also protects HUVECs from ox-LDL-induced cell damage by reducing lipid peroxidation, restoring antioxidant system activity, and decreasing cell death (Chen et al., 2010).

Del-3-glu, which is carried into endothelial cells through sodium-dependent glucose transporter, may be related with reduction of ox-LDL-induced mitochondrial dysfunctions and apoptosis, as well as decreased formation of ROS (SGLT1). Del-3-glu can reverse the effects of oxLDL-induced mitochondrial dysfunction in HUVECs when given the SGLT1 inhibitor phlorizin or SGLT1 siRNA (Jin et al., 2013). Because SGLT1 transports Del-3-glu through the cellular membrane in a temperature, concentration, and time-dependent manner, it's possible that SGLT1 regulates at least part of Del-3-glu distribution and bioavailability. However, the expression profiles of SGLT1 in different tissues may differ, making study of Del-3-glu's biological function more difficult. Anthocyanins have been studied for their cardioprotective properties by altering mitochondrial metabolism (Liobikas et al. (2016).

Del-3-glu protects against cardiac ischemia by functioning as a substrate to boost the activity of complex I of the mitochondrial respiratory chain, as seen by increased state 3 respiration (53%) and ATP generation (37%) following ischemia. Del-3-glu, which has a high reducing capacity, may act as an electron acceptor in complex I-mediated oxidation of DADH, alleviate ischemia-induced inhibition of complex I, and finally protect cardiac cells from cell death caused by energy depletion (Liobikas et al. 2016). Del-3-glu has a considerably lower affinity for complex I than CoQ1 (Skemiene et al., 2015). Del-3-glu, on the other hand, may transmit electrons to cytochrome c directly. Complex I/Del-3-glu/cytochrome c/complex IV is a novel respiratory chain that has been created. Macrophages promote atherogenesis and atherosclerosis development by internalizing lipids and transforming into foam cells, and they play a key role in plaque formation and progression.

Anthocyanins have been shown to protect against the development of atherosclerosis *via* controlling lipid buildup. Del-3-glu and its metabolites (gallic acid and syringic acid) may inhibit fat accumulation in monocytic THP1 cell-driven macrophages at low concentrations (0.050.1 g/mL) (Del Bo et al., 2016a,b). P-selectin is an adhesion molecule that mediates contact between platelets and leukocytes or endothelial cells, and it plays a key role in the progression of atherosclerosis. Del-3-glu has been shown to reduce platelet activation and P-selectin expression generated by the agonist ADP (Krga et al., 2018a,b). TNF-induced adherence of monocytes to endothelial cells is inhibited by anthocyanins extracted from blueberries. Del-3-glu, on the other hand, is proven to be ineffective for this inhibition (Del Bo et al., 2016a,b; 2019). Blood flow-induced shear stress promotes the activation of the Akt/eNOS signalling pathway in vascular endothelial cells.

Anthocyanins may change hemodynamic properties, relying more on the structure of the aglycon than the structure of the glycoside. Del-3-rut (1 mg/kg) does not appear to provide effective protection by modulating cremaster arteriole blood flow and regulating eNOS and Akt phosphorylation in the aorta (Tsutsumi et al., 2019). The endothelial vasodilator ERαis involved in the control of endothelial vasodilation. Del activates NO signalling pathways (Src, ERK1/2, eNOS, and caveolin-1) in the aorta of ERα wild-type but not ERα-KO mice, which could be associated to activation of NO signalling pathways (Src, ERK1/2, eNOS, and caveolin-1).

Virtual docking demonstrates that Del binds to ERα with a similar docking model to 17-estraliol (E2), however E2 only activates ER to a 73% capacity (Chalopin et al., 2010). Del-3-glu and Del-3-rut also bind to ERβ with IC50 values of 9.7 and 2.3 μM, respectively, (Nanashima et al., 2015, 2018), which are lower than those to ERα with IC50 values of 14.1 and 14.6 μM, respectively (Nanashima et al., 2015, 2018). Del-3-glu binds to ERβ with IC50 values of 63.2 M μM, although it does not reach the binding sites, according to another study (Hidalgo et al., 2012). Virtual docking demonstrates that there are no acceptable docking domains for glucoside or rutinoside within the ligand-binding pocket of ERβ (Nanashima et al., 2015, 2018). Furthermore, Del inhibits T lymphocyte proliferation in an ER-dependent way. Inhibition of the Ca^{2+} signalling pathway and inactivation of HDACs, NFATs, and ERK1/2 are thought to be the underlying mechanisms. Del prevents T cells from differentiating into Th1, Th17, and Treg, but it has little effect on Th2 differentiation (Dayoub et al., 2017).

In rats, anthocyanin-rich red cabbage extract (ARCE) was found to alleviate isoproterenol-induced myocardial infarction, as evidenced by enhanced caveolin-3 and SERCA2a expression, as well as reduced membrane damage and calcium dysregulation. Del-3-glu is one among the key components of ARCE, and it nearly docks with β1 adrenergic receptor (a Glide XPG score of 8.5 kcal/mol), which is the receptor for isoproterenol with high affinity calcium dependence and a Glide XPG score of 8.5 kcal/mol. 13 Del may bind with the SH2 domain of STAT1 and decrease its activity, according to virtual docking. Del pretreatment of isolated rat hearts before regional ischemia/reperfusion reduces apoptotic and necrotic cells and inhibits STAT1 activity. Del, on the other hand, had no effect on STAT3 phosphorylation levels (Scarabelli et al., 2009). The renin-angiotensin system (RAS) is important in the development of hypertension. Del has been discovered to block RAS in HEK293 cells by reducing the activity and expression of angiotensin-converting enzyme (ACE). Del also reduces the expression of renin's mRNA but has no effect on its protein expression. Del, on the other hand, was unable to bind to the angiotensin II type 1 receptor and had no influence on its internalisation (Parichatikanond et al., 2012).

3.7. Anti-Diabetic Properties of Delphinidin

Insulin-dependence and insulin-independence are two possible mechanisms of anthocyans in antidiabetes. The former entails improving the function of β-cells as well as tissue sensitivity. Reduced glucose absorption, starch breakdown enzymes, and alterations in energy metabolism are all part of the latter. Del inhibits oxidative stress, NF-κB and JNK signalling pathways activity, and protein tyrosine phosphatase 1B (PTP1B) expressions, which may help to alleviate obesity, caused by a high-fat diet and improve dyslipidemia and insulin resistance (Daveri et al., 2018). Hepatocyte nuclear factor-1α (HNF-1α) is important for maintaining glycemia and the functioning of β--cells. Del-3-ara has been demonstrated to activate HNF1αby binding to its dimerization domain and transcription domain as a homodimer with high affinity (7.2 kcal/mol). Del also has a high affinity (6.9 kcal/mol and 7.8 kcal/mol, respectively). This could mean that glycosylation of Del isn't expected to be required for interaction with HNF-1α(Real Hernandez et al., 2015).

In pancreatic cells, the free fatty acid receptor 1 (FFAR1, also known as GPR40) promote insulin secretion depending upon glucose in response to the medium to long chain fatty acids. FFAR1 helps to keep glucose levels in check by reducing glucose tolerance and fasting hyperglycemia. In diet-induced obesity mice, glucokinase (GK) has similar biological effects via lowering blood glucose levels. By interacting with the glucose transporter 2 (GLUT2) and the sodium glucose linked transporter 1, anthocyanins, notably Del-3-glu, have been demonstrated to inhibit epithelial glucose transfer (SGLT1). Furthermore, Del-3-glu (100 μM) greatly boosts the expression of FFAR1, a G protein-subunit of the Gq family that is linked to G protein. Phospholipase C (PLC) and phosphorylated protein kinase D (PPKD) are likewise upregulated by Del-3-glu (p-PKD). Furthermore, Del-3-glu increases glucose absorption in hepatic cells by upregulating GK and downregulating phosphorylated AMPK and phosphoenolpyruvate carboxy kinase (PEPCK) activity (Luna Vital and Gonzalez de Mejia, 2018). Del also boosts the activity of FFAR1 and FFAR1-mediated intracellular calcium release and calcium oscillations in human intestinal cell lines, leading to increased cAMP activity and glucose absorption (Hidalgo et al., 2017).

Del-3-rut is the primary ingredient in black currant polyphenolic extracts, which has been demonstrated to lower blood glucose levels and enhance glucose tolerance in type II diabetic rats via boosting AMPK/Glut4 signalling and glucagon-like peptide-1 (GLP-1) activity (Iizuka et al., 2018). Del-3-rut maintains its structure after 45-60 minutes of ingestion and enhances insulin production by stimulating GLP-1 expression, according to a new study. These findings suggest that Del-3-rut, rather than its breakdown products, is primarily responsible for GLP-1 secretion (Tani et al., 2017). Del-3-rut's role in modulating GLP-1 secretion could be linked to the stimulation of the GRP40/120/Ca^{2+}-CaMKII signalling pathway in GLUTag cells (Kato et al., 2015). Dipeptidyl peptidase IV (DPP-IV), a serine aminopeptidase and a target for type 2 diabetes therapy, rapidly degrades GLP-1. Del-3-ara increases insulin secretion and expression of insulin-receptor associated factors in pancreatic β-cells by potentiating the activities of DPP-IV and GLP-1 (Johnson and De Mejia, 2016). Starch-degrading enzymes (α-glucosidase and α-amylase) play a key role in blood glucose elevation. Del-3-glu has recently been demonstrated to have potential activity against α-glucosidase and α-amylase, which is connected to Del-3-glu's high hydroxyl group content and polarity (Osteberg Potthoff et al., 2019).

Del (100 μM) inhibits the activities of α-glucosidase, α-amylase, and DPP-IV in Caco-2 cells and reduces ROS generation. Del-3-glu has a moderate inhibitory impact on these three enzymes, with free energy values of −4.77, −7.35, and −5.00kcal/mol, respectively (Mojica et al., 2017).Similarly, Del-3-glu is one of the main compounds found in black soybean seed coat extract, which reduces the activity of α-glucosidase and α-amylase, inhibits lipid accumulation, and reduces H_2O_2-induced oxidative stress in HepG2 cells, resulting in hypoglycemia and hypolipidemia in type 2 diabetes mellitus (Chen et al., 2018a,b). DIAVIT is a natural medicine made from polyphenol-rich extracts of blueberries and sea buckthorn, both of which are high in Del. DIAVIT has been proven to protect db/db mice from developing type 2 diabetic nephropathy, as seen by improved hyperglycemia and albuminuria, as well as increased glomerular water permeability. In podocytes *in vitro*, DIAVIT also efficiently inhibits aberrant VEGF-A splicing and enhances the ratio of VEGF-A165b/total VEGF-A165 (Stevens et al., 2019).

3.8. Anti-Obesity Activity of Delphinidin

Obesity is a global epidemic health concern because of its risks for developing various complications, including cardiovascular diseases, diabetes, and cancers. The black soybean anthocyanins containing Del-3-glu have been shown to reduce the viability, proliferation, and differentiation of 3T3-L1 cells by decreasing the expression of PPARγ (Kim et al., 2012a,b). Cellular energy balance has been linked to the AMPK-mediated signalling system. Del-3-glu has been shown to boost AMPK and acetyl-CoA carboxylase (ACC) phosphorylation and decrease PPARγ, C/EBPα, SREBP1, and FAS expression, lowering lipid levels and tiny lipid droplets, attenuating adipogenesis, and promoting lipid metabolism (Park et al., 2019). Del inhibits intracellular lipid accumulation and increases lipolysis in early 3T3-L1 preadipocytes by inducing G1 cell cycle arrest and decreased expression of transcription factors governing early adipogenesis, which may be linked to activation of the Wnt/β-catenin signalling pathway (Rahman et al., 2016). SMARCD1 inhibition has been linked to palmitic acid-induced lipid accumulation (PA). Del-3-glu suppresses lipid accumulation and cell senescence induced by PA in hepatocytes by restoring SMARCD1 expression (Harada et al., 2018).

Del has been shown to inhibit MSC adipogenesis and enhance chondrogenesis, as evidenced by decreased expression of FABP4 and adiponectin and increased expression of Col2a1 and aggrecan (Saulite et al., 2019). Del inhibits the accumulation of triglycerides and the expression of CPT1A, SREBF1, and FASN in HepG2 cells without changing the activity of AMPK. Del, surprisingly, had no effect on body weight, hyperglycemia, insulin resistance, or histochemistry as a result of high fat or high-carbohydrate meals (Parra Vargas et al., 2018). Anthocyanins, such as cyanidin-3-glucoside, have been shown to ameliorate fatty liver caused by a high-fat, high-cholesterol (HFC) diet by downregulating eNAMPT and FGF21 expression while upregulating NRG4 and PGC-1αexpression. Unfortunately, in HFC diet-fed mice treated with cyanidin-3-glucoside, body weight and BAT index do not alter significantly (Pei et al., 2018). Surprisingly, treatment with a mixture of cyanidin and Del showed lowered body weight and increased brown fat weight in mice. Treatment with a mixture of cyanidin and Del resulted in lower body weight and increased brown fat weight in mice, which was controversial (Daveri et al., 2018). This conundrum could be linked to the many animal models. As a result, greater research into the effects of anthocyanins, particularly Del and its glycosides, on obesity management *in vivo* and *in vitro* is needed.

3.9. Osteogenic Activity of Delphinidin

Delphinidin suppressed osteoclastogenesis and osteoclastogenic transcription factors in RAW 264.7 macrophages, according to (Moriwaki et al., 2014). Delphinidin-3-rutinoside also boosted the expression of osteoblastogenic markers in MC3T3-E1 cells, according to (Casati et al., 2019). Malvidin increased calcium buildup in osteocytes and osteogenic transcription factor expression (Saulite et al., 2019). In RAW 264.7 macrophages, cyanidin chloride treatment reduced osteoclastogenesis and osteoclast activity (Cheng et al., 2018). However, whereas a large dose of cyanidin reduced osteoclastogenesis, a modest amount encouraged it, according to (Dou et al.,

2016). In addition, cyanidin, like delphinidin, inhibited calcium buildup in osteocytes (Saulite et al., 2019).

Based on these findings, *in vitro* treatment with the separated anthocyanin chemicals delphinidin and malvidin appears to promote bone formation, although the effects of cyanidin remain unknown (Sakaki et al., 2020). One of the flavonoid sub-type, anthocyanin is vast family of water-soluble phytopigments with a variety of health-promoting properties. In ovariectomized (OVX) animal models, an anthocyanin-rich molecule from blueberries was recently found to have preventive properties against bone loss. The active elements in the anthocyanin complex, however, have yet to be discovered. Authors showed that delphinidin, one of the major anthocyanidins found in berries, is a powerful anti-osteoporotic active component that works by preventing bone resorption. When compared to other anthocyanidins, such as cyanidin and peonidin, delphinidin administration significantly prevented the development of RAW264.7 cells into osteoclasts *in vitro*. In both RANKL-induced osteoporosis model mice and OVX model mice, oral treatment of delphinidin dramatically reduced bone loss. We also found that delphinidin inhibited the activity of the major transcriptional factors for osteoclastogenesis, NF-κB, c-Fos, and Nfatc1. These findings strongly suggest that delphinidin is the most potent osteoclast differentiation inhibitor and will be an excellent treatment for postmenopausal osteoporosis (Moriwaki et al., 2014).

3.9.1. Inhibitory Effects of Delphinidin on Bone Loss

Authors explored the anti-osteoporotic impact of delphinidin on sRANKL-induced osteoporosis model mice to establish the applicability of our *in vitro* results of delphinidin bioactivity to the *in vivo* scenario. When compared to intact control mice, 3D pictures and morphometric metrics clearly showed that sRANKL therapy significantly reduced the cancellous bone volume of the femoral distal metaphysis. Delphinidin, on the other hand, substantially reduced cancellous bone deterioration in sRANKL-treated animals. For 17 days, all animals were weighed daily, and there was no significant difference between the experimental groups. Although sRANKL-treated animals had slightly lower liver and kidney weights than control mice, the changes were not statistically significant. Furthermore, no difference was seen between sRANKL-treated mice and sRANKL/delphinidin-treated animals (Moriwaki et al., 2014). Authors then examined the bone tissues histomorphometrically and found that the number of bone trabeculare was clearly reduced in sRANKL-treated osteoporotic mice, according to microscopic examination.

Many bone trabeculae, on the other hand, persisted in the bones of delphinidin-treated mice. Despite the restricted number of bone trabeulare, the number of osteoclasts in delphinidin-treated mice was considerably lower ($p= 0.03$) than in non-treated osteoporotic mice.

Other two significant resorption-parameters were similarly down-regulated in mice given delphinidin; however no statistically significant changes were seen between these two groups. Consistent with the in vitro findings, these findings showed that delphinidin works directly on osteoclast differentiation (Moriwaki et al., 2014). We then tested delphinidin's anti-bone resorption activity in OVX mice, a classic osteoporosis model. Three different dosages of delphinidin were given orally to OVX mice. Over half of the mice treated with not only high-dose (10 mg/kg) but also intermediate-dose (3 mg/kg) delphinidin showed a substantial anti-osteoporotic effect. In this study, uterine weight did not alter in delphinidin-treated OVX mice (data not shown), indicating that delphinidin did not act on the uterus in the same way as SERMs (selective oestrogen receptor modulators) like isoflavons (Diel et al., 2004). Our findings show that ingesting delphinidin serves as a strong anti-osteoporotic drug in the prevention of osteoporosis.

3.9.2. Mechanism of Delphinidin-Mediated Inhibition of Osteoclastogenesis

Because earlier research had shown that anthocyanin extract reduced NF-B activation in activated monocytes, we initially used the TransAM test to look at NF-B activity in RAW264.7 cells. When the cells were stimulated with sRANKL, the activation of the NF-B p65 subunit was immediately increased. In contrast, RANKL-mediated NF-B activation was significantly reduced in cells treated with delphinidin, indicating a transcriptional impact. Another critical event in the early stages of osteoclastogenesis is RANKL-induced c-Fos activation (Takayamagi et al., 2002; Wada et al., 2006). In osteoclastogenesis, NFATc1 is a master transcriptional factor found at the downstream crossroads of the NF-B and c-Fos pathways (Takayamagi et al., 2002; Wada et al., 2006). Authors wanted to find if delphinidin impacts the activation of the c-Fos and Nfatc1 genes. In addition, they looked at how delphinidin affected the expression of osteoclast marker genes in RAW264.7 cells. Our findings indicate that delphinidin therapy reduced RANKL-induced c-Fos activation. Furthermore, Nfatc1 gene expression was significantly reduced in delphinidin-treated cells. As a result, delphinidin therapy decreased the expression of osteoclast marker genes, particularly Mmp9 and Trap. Furthermore, the expression of Dc-stamp, a fusion protein gene for multinucleation, was reduced significantly in delphinidin-treated cells. These findings show that the anti-osteoclastogenic action of delphinidin is obviously linked to a change in the signalling cascade triggered by RANKL.

3.10. Impact of Delphinidin on Other Metabolic Pathways

The effect of delphinidin on human Glyoxalase I (GLO I) was explored, and it was found to have the most potent inhibitory effect on human Glyoxalase I (GLO I), indicating that it could be used to produce new GLO I inhibitory anticancer medicines (Takasawa et al., 2010). Delphinidin has a redox-independent effect on topoisomerase II alpha catalytic activity. In HT29 cells, delphinidin reduced the DNA-damaging effects of topoisomerase II poisons without changing the level of sites sensitive to formamidopyrimidine-DNA-glycosylase (Esselen et al., 2009). In a sensitive and robust fluorometric high-throughput screening assay, delphinidin was reported to have inhibitory capability against *Schistosoma mansoni* NAD (+) catabolizing enzyme (SmNACE) (Kuhn et al., 2010). In another investigation, Delphinidin reduced TNF-alpha-induced Cox-2 expression in JB6 P+ mouse epidermal (JB6 P+) cells. TNF-alpha-induced phosphorylation of JNK, p38 MAP kinase, Akt, p90RSK, MSK1, and ERK was suppressed by Delphinidin, which then hindered the activation of the eukaryotic transcription factors AP-1 and NF-kappaB (Hwang et al., 2009).

The effect of delphinidin (Dp) on soybean lipoxygenase-1 and human neutrophil granulocyte 5-lipoxygenase was studied, and it was discovered that Dp 3-O-glucoside (Dp3glc) and Dp 3-O-galactoside (Dp3gal) inhibited soybean lipoxygenase-1 the most effectively (Knaup et al., 2009). The endothelium-dependent vasorelaxation action of delphinidin was studied in mice lacking the alpha isoform of the oestrogen receptor (ERalpha). Delphinidin activates the nitric oxide (NO) pathway in endothelial cells, causing endothelial vasodilation in the aorta of ERalpha Wild-Type but not Knock-Out mice (Chalopin et al., 2010). Another study (Korte et al., 2009) used competitive radioligand binding experiments to identify delphinidin as ligands with intermediate affinity for human cannabinoid receptor 1.

Based on the another study used dye efflux, ATPase, and, for BCRP, vesicular transport tests to expose anthocyanins and anthocyanidins to the human efflux transporters multidrug resistance protein 1 (MDR1) and breast cancer resistance protein (BCRP). The BCRP transporter was found to interact with delphinidin (Dreiseitel et al., 2009a,b). At low micromolar doses, delphinidin was reported to substantially block the protein tyrosine kinase activity of receptor tyrosine kinases (Teller et al., 2009).A photometric assay was utilized in another work to test delphinidin's inhibitory effectiveness against the phospholipase enzyme (Dreiseitel et al., 2009a,b). The antioxidant and antiangiogenic effects of delphinidin from *Vaccinium myrtillus* were examined. In a co-culture of human umbilical vein endothelial cells (HUVECs) and fibroblasts, these anthocyanidins reduced VEGF-induced tube formation in a concentration-dependent manner (Matsunaga et al., 2010). The effect of delphinidin as a spermidine-induced proPlasma hyaluronan-binding protein (PHBP) autoactivation inhibitor was examined, and it was found to be potent and selective, while heparin-induced pro-PHBP was not inhibited (Yamamoto et al., 2011).

Delphinidin chloride was found to be unstable in Dulbecco's modified Eagle's medium (DMEM), RPMI 1640 (RPMI), and Minimal Essential Medium Eagle (MEM) culture conditions in another investigation (Long et al., 2010). The effect of red wine extract on plasma antioxidant status and monocyte chemoattractant protein 1 production in healthy human subjects was examined by looking at the absorption and metabolism of the 3-monoglucosides of delphinidin in healthy human individuals (Garcia Alonso et al., 2009). Gene E1 is involved in the manufacture of trihydroxylated delphinidin and is postulated as a scheme of genetic control of anthocyanidin biosynthesis in sweet pea flowers in another study (Rat'kin and Tarasov, 2010). Delphinidin was discovered to have a high interaction with soybean seed ferritin in another investigation (Deng et al., 2010).

Conclusion and Future Perspectives of Delphinidin

Nature is important in our lives because it provides all of the raw materials for food, clothing, and even drugs used to treat various diseases. To maintain proper growth, the pharmaceutical industries must innovate and gain access to high output rates on low-cost materials while maintaining reasonable safety. The combination of advanced chemistry with bio-based starting materials has the potential to revolutionise the pharmaceutical industry. Due to its important contribution to human health, there should be a focus on increasing production and improving delphinidin content to satisfy industrial demand. There is still potential for growth in strategies for extracting larger levels of delphinidin while lowering its cost. Tissue culture techniques may be the best choice in the future for producing the maximum level of delphinidin. Delphinidin plays an important role in human health; therefore further research into its general health benefits, including its usage as a nutraceutical and dietary supplement, should be conducted in the future. However, additional research should be done to evaluate its safety profile before it is used. In terms of its safety profile, case study data should be collected to learn about its impact on humans and other creatures. In this regard, both preclinical and clinical studies are needed to investigate the hidden potential of delphinidin in the treatment of different illnesses.

References

Aggarwal, D., Upadhyay, S. K., Singh, R. & Tuli, H. S. (2021). Recent patents on therapeutic activities of xanthohumol: A prenylated chalconoid from hops (*Humulus lupulus* L.). *Pharmaceutical Patent Analyst*, 10(1), 37–49.

Aura, A. M., Martin-Lopez, P., O'Leary, K. A., Williamson, G., Oksman-Caldentey, K. M., Poutanen, K. & Santos Buelga, C. (2005). In vitro metabolism of anthocyanins by human gut microflora. *European Journal of Nutrition,* 44, 133–142.

Bae, C. H., Jeon, B. S., Choi, Y. S., Song, S. Y. & Kim, Y. D. (2014). Delphinidin inhibits Lps-induced Muc8 and Muc5b expression through Toll-Like Receptor 4-Mediated Erk1/2 and P38 Mapk in human airway epithelial cells. *Clinical and Experimental Otorhinolaryngology,* 7(3), 198−204.

Bhagwat, S., Haytowitz, D. B., Wasswa-Kintu, S. I. & Holden, J. M. (2013). USDA develops a database for flavonoids to assess dietary intakes. *36th National Nutrient Databank Conference*, 2, 81–86.

Bin Hafeez, B., Asim, M., Siddiqui, I. A., Adhami, V. M., Murtaza, I. & Mukhtar, H. (2008). Delphinidin, a dietary anthocyanidin in pigmented fruits and vegetables: A new weapon to blunt prostate cancer growth. *Cell Cycle,* 7, 3320–3326.

Bishayee, A., Háznagy-Radnai, E., Mbimba, T., Sipos, P., Morazzoni, P., Darvesh, A. S., Bhatia, D. & Hohmann, J. (2010). Anthocyanin-rich black currant extract suppresses the growth of human hepatocellular carcinoma cells. *Natural Product Communications*,5(10), 1613–1618.

Bonesi, M., Loizzo, M. R., Menichini, F. & Tundis, R. (2018). Immunity and inflammation in health and disease. II. Flavonoids in treating psoriasis. *Academic Press, London*, pp. 281–294.

Bunea, A., Rugină, D., Sconţa, Z., Pop, R. M., Pintea, A., Socaciu, C., Tăbăran, F., Grootaert, C., Struijs, K. & Van-Camp, J. (2013). Anthocyanin determination in blueberry extracts from various cultivars and their antiproliferative and apoptotic properties in B16-F10 metastatic murine melanoma cells. *Phytochemistry,* 95, 436–444.

Calderón-Oliver, M. & Ponce-Alquicira, E. (2018).Fruits: A source of polyphenols and health benefits. *Natural and Artificial Flavoring Agents and Food Dyes,* pp. 189–228.

Casati, L., Pagani, F., Fibiani, M., Lo Scalzo, R. & Sibilia, V. (2019). Potential of delphinidin-3-rutinoside extracted from *Solanum melongena* L. as promoter of osteoblastic MC3T3-E1 function and antagonist of oxidative damage. *European Journal of Nutrition*, 58(3), 1019−1032.

Chalopin, M., Tesse, A., Martinez, M. C., Rognan, D., Arnal, J. F. & Andriantsitohaina, R. (2010). Estrogen receptor alpha as a key target of red wine polyphenols action on the endothelium. *PLoS One*, 5(1), https://doi.org/10.1371/journal.pone.0008554.

Charepalli, V., Reddivari, L., Vadde, R., Walia, S., Radhakrishnan, S. & Vanamala, J.K.P. (2016). *Eugenia jambolana* (Java plum) fruit extract exhibits anti-cancer activity against early stage human HCT-116 colon cancer cells and colon cancer stem cells. *Cancers,* 8(3), 29. https://doi.org/10.3390/cancers8030029.

Chen, C. Y., Yi, L., Jin, X., Zhang, T., Fu, Y. J., Zhu, J. D., Mi, M. T., Zhang, Q. Y., Ling, W. H. & Yu, B. (2011). Inhibitory effect of delphinidin on monocyte-endothelial cell adhesion induced by oxidized low-density lipoprotein via Ros/P38mapk/Nf-Kappab pathway. *Cell Biochemistry and Biophysical*, 61(2), 337−348.

Chen, C.Y., Yi, L., Jin, X., Mi, M.T., Zhang, T., Ling, W.H. &Yu, B. (2010). Delphinidin attenuates stress injury induced by oxidized lowdensity lipoprotein in human umbilical vein endothelial cells. *Chemico-Biological Interaction*, 183, 105–112.

Chen, Z., Wang, C., Pan, Y., Gao, X. & Chen, H. (2018a). Hypoglycemic and hypolipidemic effects of anthocyanins extract from black soybean seed coat in high fat diet and streptozotocin-induced diabetic mice. *Food and Function*, 9(1), 426−439.

Cheng, J., Zhou, L., Liu, Q., Tickner, J., Tan, Z., Li, X., Liu, M., Lin, X., Wang, T., Pavlos, N. J., Zhao, J. & Xu, J. (2018b). Cyanidin chloride inhibits ovariectomy-induced osteoporosis by suppressing RANKL-mediated osteoclastogenesis and associated signaling pathways. *Journal of Cell Physiology*, 233(3), 2502–2512.Chen, J., Li, H. Y., Wang, D. & Guo, X. Z. (2019a). Delphinidin protects Beta2m-/Thy1+ bone marrow-derived hepatocyte stem cells against Tgf-Beta1-induced oxidative stress and apoptosis through the Pi3k/Akt pathway in vitro. *Chemico-Biological Interactions*, 297, 109−118.

Chen, Z., Zhang, R., Shi, W., Li, L., Liu, H., Liu, Z. & Wu, L. (2019b). The multifunctional benefits of natural occurring delphinidin and its glycosides. *Journal of Agricultural and Food Chemistry,* 67(41), 11288–11306.

Cho, J. S., Kang, J. H., Shin, J. M., Park, I. H. & Lee, H. M. (2015). Inhibitory effect of delphinidin on extracellular matrix production via the Mapk/Nf-Kappab pathway in nasal polyp-derived fibroblasts. *Allergy, Asthma Immunology Research*, 7(3), 276−282.

Chorfa, N., Savard, S. & Belkacemi, K. (2016). An efficient method for high-purity anthocyanin isomers isolation from wild blueberries and their radical scavenging activity. *Food Chemistry,* 197(Pt B), 1226−1234.

Clere, N., Faure, S., Martinez, M. C. & Andriantsitohaina, R. (2011). Anticancer properties of flavonoids: Roles in various stages of carcinogenesis. *Cardiovascular and Hematological Agents in Medical Chemistry*, 9, 62–77.

Cremonini, E., Mastaloudis, A., Hester, S. N., Verstraeten, S. V., Anderson, M., Wood, S. M., Waterhouse, A. L., Fraga, C. G. & Oteiza, P. I. (2017). Anthocyanins Inhibit tumor necrosis alpha-induced loss of Caco-2 cell barrier integrity. *Food and Function,* 8(8), 2915–2923.

Cvorovic, J., Tramer, F., Granzotto, M., Candussio, L, Decorti, G. & Passamonti, S. (2010). Oxidative stress-based cytotoxicity of delphinidin and cyanidin in colon cancer cells. *Archives of Biochemistry and Biophysics*, 501, 151–157.

Daveri, E., Cremonini, E., Mastaloudis, A., Hester, S. N., Wood, S. M., Waterhouse, A. L., Anderson, M., Fraga, C. G. & Oteiza, P. I. (2018). Cyanidin and delphinidin modulate inflammation and altered redox signaling improving insulin resistance in high fat-fed mice. *Redox Biology*, 18, 16–24.

Dayoub, O., Andriantsitohaina, R. & Clere, N. (2014). Pleiotropic beneficial effects of epigallocatechin gallate, quercetin and delphinidin on cardiovascular diseases associated with endothelial dysfunction. *Cardiovascular and Hematological Agents in Medical Chemistry*, 11(4), 249–264.

Dayoub, O., Le Lay, S., Soleti, R., Clere, N., Hilairet, G., Dubois, S., Gagnadoux, F., Boursier, J., Martinez, M. C. and Andriantsitohaina, R. (2017). Estrogen receptor Alpha/Hdac/Nfat axis for delphinidin effects on proliferation and differentiation of T lymphocytes from patients with cardiovascular risks. *Science Report*, 7(1), 937–938.

Del Bo', C., Cao, Y., Roursgaard, M., Riso, P., Porrini, M., Loft, S. & Møller, P. (2016a). Anthocyanins and phenolic acids from a wild blueberry (*Vaccinium angustifolium*) powder counteract lipid accumulation in Thp-1-derived macrophages. *European Journal of Nutrition*, 55(1), 171–182.

Del Bo', C., Roursgaard, M., Porrini, M., Loft, S., Møller, P. & Riso, P. (2016b). Different effects of anthocyanins and phenolic acids from wild blueberry (*Vaccinium angustifolium*) on monocytes adhesion to endothelial cells in a Tnf-alpha stimulated proinflammatory environment. *Molecular Nutrition and Food Research*, 60(11), 2355–2366.

Del Bo', C., Marino, M., Riso, P., Møller, P. & Porrini, M. (2019). Anthocyanins and metabolites resolve Tnf-Alpha-mediated production of E-selectin and adhesion of monocytes to endothelial cells. *Chemico-Biological Interactions*, 300, 49–55.

Deng, J., Cheng, J., Liao, X., Zhang, T., Leng, X. & Zhao, G. (2010). Comparative study on iron release from soybean (*Glycine max*) seed ferritin induced by anthocyanins and ascorbate. *Journal of Agricultural and Food Chemistry*, 58, 635–641.

Diaconeasa, Z., Leopold, L., Rugină, D., Ayvaz, H. & Socaciu, C. (2015). Antiproliferative and antioxidant properties of anthocyanin rich extracts from blueberry and black currant juice. *International Journal of Molecular Science,* 16, 2352–2365.

Diel P, Geis RB, Caldarelli A, Schmidt S, Leschowsky UL, Voss, A. & Vollmer, G. (2004). The differential ability of the phytoestrogen genistein and of estradiol to induce uterine weight and proliferation in the rat is associated with a substance specific modulation of uterine gene expression. *Molecular and Cellular Endocrinology*, 221(1&2), 21–32.

Dimitric Markovic, J. M., Pejin, B., Milenkovic, D., Amic, D., Begovic, N., Mojovic, M. & Markovic, Z. S. (2017). Antiradical activity of delphinidin, pelargonidin and malvin towards hydroxyl and nitric oxide radicals: The energy requirements calculations as a prediction of the possible antiradical mechanisms. *Food Chemistry,* 218, 440–446.

Domitrović, R. & Jakovac, H. (2010). Antifibrotic activity of anthocyanidin delphinidin in carbon tetrachloride-induced hepatotoxicity in mice. *Toxicology,* 4, 1–10.

Dou, C., Li. J., Kang, F., Cao, Z., Yang, X., Jiang, H., Yang, B., Xiang, J., Xu, J. & Dong, S. (2016). Dual effect of cyanidin on RANKL-induced differentiation and fusion of osteoclasts. *Journal of Cell Physiology*, 231(3), 558–567.

Dreiseitel, A., Korte, G., Schreier, P., Oehme, A., Locher, S., Hajak, G. & Sand, P. G. (2009a). Phospholipase A(2) is inhibited by anthocyanidins. *Journal of Neural Transmission*, 116, 1071–1077.

Dreiseitel, A., Oosterhuis, B., Vukman, K.V., Schreier, P., Oehme, A., Locher, S., Hajak, G. & Sand P. G. (2009b). Berry anthocyanins and anthocyanidins exhibit distinct affinities for the efflux transporters BCRP and MDR1. *British Journal of Pharmacology*, 158, 1942–1950.

Edirisinghe, I., Banaszewski, K., Cappozzo, J., McCarthy, D. & Burton-Freeman, B. M. (2011). Effect of black currant anthocyanins on the activation of endothelial nitric oxide synthase (Enos) in vitro in human endothelial cells. *Journal of Agricultural and Food Chemistry,* 59(16), 8616–8624.

Esselen, M., Fritz, J., Hutter, M. & Marko, D. (2009). Delphinidin modulates the DNA-damaging properties of topoisomerase II poisons. *Chemical Research in Toxicology,* 22, 554–564.

Estévez, L. & Mosquera, R.A. (2008). Molecular structure and antioxidant properties of delphinidin. *The Journal of Physical Chemistry-A*, 112, 10614–10623.

Fang, J. (2015). Classification of fruits based on anthocyanin types and relevance to their health effects. *Nutrition,* 31, 1301–1306.

Feng, R., Wang, S. Y., Shi, Y. H., Fan, J. & Yin, X. M. (2010). Delphinidin induces necrosis in hepatocellular carcinoma cells in the presence of 3-methyladenine, an autophagy inhibitor. *Journal of Agricultural and Food Chemistry*, 58, 3957–3964.

Flores, G., Dastmalchi, K., Paulino, S., Whalen, K., Dabo, A. J., Reynertson, K. A., Foronjy, R. F., D'Armiento, J. M. & Kennelly, E. J. (2012). Anthocyanins from *Eugenia brasiliensis* edible fruits as potential therapeutics for copd treatment. *Food Chemistry,* 134(3), 1256–1262.

Fritz, J., Roth, M., Holbach, P., Esselen, M. & Marko, D. (2008). Impact of delphinidin on the maintenance of DNA integrity in human colon carcinoma cells. *Journal of Agricultural Food Chemistry*, 56, 8891–8896.

Garbacki, N., Angenot, L., Bassleer, C., Damas, J. & Tits, M. (2002). Effects of prodelphinidins isolated from *Ribes nigrum* on chondrocyte metabolism and cox activity. *Naunyn-Schmiedeberg's Archives of Pharmacology,* 365(6), 434–441.

Garcia-Alonso, M., Minihane, A. M., Rimbach, G., Rivas-Gonzalo, J. C. & de Pascual-Teresa, S. (2009). Red wine anthocyanins are rapidly absorbed in humans and affect monocyte chemoattractant protein 1 levels and antioxidant capacity of plasma. *Journal of Nutritional Biochemistry*, 20, 521–529.

Gharib, A., Faezizadeh, Z. & Godarzee, M. (2013). Treatment of diabetes in the mouse model by delphinidin and cyanidin hydrochloride in free and liposomal forms. *Planta Medica*, 79, 1599–1604.

Goszcz, K., Deakin, S. J., Duthie, G. G., Stewart, D. & Megson, I. L. (2017). Bioavailable concentrations of delphinidin and its metabolite, gallic acid, induce antioxidant protection associated with increased intracellular glutathione in cultured endothelial cells. *Oxidative Medicine and Cellular Longevity*, https://doi.org/10.1155/2017/9260701.

Grumezescu, A. M. & Holban, A. M. (2017). Natural and artificial flavoring agents and food dyes. *In:* Handbook of food bioengineering, technology and engineering.Routledge Taylor and Francis*CRC Press*,7, 566p.

Hafeez, B. B., Siddiqui, I. A., Asim, M., Malik, A., Afaq, F., Adhami, V. M., Saleem, M., Din, M. & Mukhtar, H. (2008). A dietary anthocyanidin delphinidin induces apoptosis of human prostate cancer PC3 cells in vitro and in vivo: Involvement of nuclear factor-kappaB signaling. *Cancer Research*, 68, 8564–8572.

Harada, G., Onoue, S., Inoue, C., Hanada, S. & Katakura, Y. (2018). Delphinidin-3-glucoside suppresses lipid accumulation in Hepg2 cells. *Cytotechnology*, 70(6), 1707−1712.

Hariri, B. M., Payne, S. J., Chen, B., Mansfield, C., Doghramji, L. J., Adappa, N. D., Palmer, J. N., Kennedy, D. W., Niv, M. Y. & Lee, R. J. (2016). In vitro effects of anthocyanidins on sinonasal epithelial nitric oxide production and bacterial physiology. *American Journal of Rhinology and Allergy*, 30(4), 261−268.

Haseeb, A., Chen, D. & Haqqi, T. M. (2013). Delphinidin inhibits Il1beta-induced activation of Nf-Kappab by modulating the phosphorylation of Irak-1(Ser376) in human articular chondrocytes. *Rheumatology*, 52(6), 998−1008.

Hidalgo, J., Teuber, S., Morera, F. J., Ojeda, C., Flores, C. A., Hidalgo, M. A., Nunez, L., Villalobos, C. & Burgos, R. A. (2017). Delphinidin reduces glucose uptake in mice jejunal tissue and human intestinal cells lines through Ffa1/Gpr40. *International Journal of Molecular Sciences*, 18(4), 750–768.

Hidalgo, M., Martin-Santamaria, S., Recio, I., Sanchez-Moreno, C., de Pascual-Teresa, B., Rimbach, G. & de Pascual-Teresa, S. (2012). Potential anti-inflammatory, anti-adhesive, anti/estrogenic, and angiotensin-converting enzyme inhibitory activities of anthocyanins and their gut metabolites. *Genes Nutrition*, 7(2), 295−306.

Huang, W. Y., Wang, J., Liu, Y. M., Zheng, Q. S. & Li, C. Y. (2014). Inhibitory effect of malvidin on TNF-α-induced inflammatory response in endothelial cells. *European Journal of Pharmacology,* 723, 67–72.

Hwang, M. K., Kang, N. J., Heo, Y. S., Lee, K. W. & Lee, H. J.(2009). Fyn kinase is a direct molecular target of delphinidin for the inhibition of cyclooxygenase-2 expression induced by tumor necrosis factor-alpha. *Biochemical Pharmacological,* 77(7), 1213−1222.

Iizuka, Y., Ozeki, A., Tani, T. & Tsuda, T. (2018). Blackcurrant extract ameliorates hyperglycemia in type 2 diabetic mice in association with increased basal secretion of glucagon-like peptide-1 and activation of amp-activated protein kinase. *JournalofNutritionalScienceandVitaminology*, 64(4), 258−264.

Inoue, H., Maeda-Yamamoto, M., Nesumi, A. & Murakami, A. (2012). Delphinidin-3-O-galactoside protects mouse hepatocytes from (−)-epigallocatechin-3-gallate-induced cytotoxicity via up-regulation of heme oxygenase-1 and heat shock protein 70. *NutritionResearch*, 32(5), 357−364.

Isaak, C. K., Petkau, J. C., Blewett, H., Siow, O. K. & Lingonberry, Y. L. (2017). Anthocyanins protect cardiac cells from oxidative stress-induced apoptosis. *Canadian Journal of Physiology and Pharmacology*, 95(8), 904−910.

Jana, S., Patel, D., Patel, S., Upadhyay, K., Thadani, J., Mandal, R., Das, S. & Devkar, R. (2017). Anthocyanin rich extract of *Brassica oleracea* L. alleviates experimentally induced myocardial infarction. *PLoS One*, 12(8), e0182137. https://doi.org/10.1371/journal.pone.0182137.

Jara, E., Hidalgo, M. A., Hancke, J. L., Hidalgo, A. I., Brauchi, S., Nunez, L., Villalobos, C. & Burgos, R. A. (2014). Delphinidin activates Nfat and induces Il-2 production through soce in T cells. *Cell Biochemistry and Biophysics,* 68(3), 497−509.

Jiménez, A., Lisa-Santamaría, P., García-Marino, M., Escribano-Bailón, M. T., Rivas-Gonzalo, J. C. & Revuelta, J. L. (2010). The biological activity of the wine anthocyanins delphinidin and petunidin is mediated through Msn2 and Msn4 in Saccharomyces cerevisiae.*FEMS Yeast Research*,10(7), 858–869.

Jin, X., Chen, M., Yi, L., Chang, H., Zhang, T., Wang, L., Ma, W., Peng, X., Zhou, Y. & Mi, M. (2014). Delphinidin-3-glucoside protects human umbilical vein endothelial cells against oxidized low density lipoprotein-induced injury by autophagy upregulation via the Ampk/Sirt1 signaling pathway. *Molecules Nutrition and Food Research*, 58(10), 1941−1951.

Jin, X., Yi, L., Chen, M. L., Chen, C. Y., Chang, H., Zhang, T., Wang, L., Zhu, J. D., Zhang, Q. Y. & Mi, M. T. (2013). Delphinidin-3- glucoside protects against oxidized low-density lipoprotein induced mitochondrial dysfunction in vascular endothelial cells via the sodium-dependent glucose transporter Sglt1. *PLoS One*, 8(7), e68617. https://doi.org/10.1371/journal.pone.0068617.

Jing, P., Qian, B., Zhao, S., Qi, X., Ye, L., Monica Giusti, M. & Wang, X. (2015). Effect of glycosylation patterns of chinese eggplant anthocyanins and other derivatives on antioxidant effectiveness in human colon cell lines. *Food Chemistry*. 172, 183−189.

Johnson, M. H. & De Mejia, E. G. (2016). Phenolic compounds from fermented berry beverages modulated gene and protein expression to increase insulin secretion from pancreatic β-cells in vitro. *Journal of Agricultural and Food Chemistry,* 64(12), 2569–2581.

Kang, N. J., Lee, K. W., Kwon, J. Y., Hwang, M. K., Rogozin, E. A., Heo, Y. S., Bode, A. M., Lee, H. J. & Dong, Z. (2008). Delphinidin attenuates neoplastic transformation in JB6 Cl41 mouse epidermal cells by blocking Raf/mitogen-activated protein kinase kinase/extracellular signal-regulated kinase signaling. *Frontiers in Cancer Prevention Research*, 1(7), 522–531.

Kato, M., Tani, T., Terahara, N. & Tsuda, T. (2015). The anthocyanin delphinidin 3-rutinoside stimulates glucagon-like peptide-1 secretion in murine GLUTag cell line via the Ca2+/calmodulin dependent kinase ii pathway. *PLoS One*, 10(5), e0126157. https://doi.org/10.1371/journal.pone.0126157

Kausar, H., Jeyabalan, J., Aqil, F., Chabba, D., Sidana, J., Singh, I. P. & Gupta, R. C. (2012). Berry anthocyanidins synergistically suppress growth and invasive potential of human non-small-cell lung cancer cells. *Cancer Letter,* 325, 54–62.

Keppler, K. & Humpf, H. U. (2005). Metabolism of anthocyanins and their phenolic degradation products by the intestinal microflora. *Bioorganic and Medicinal Chemistry*, 13, 5195–5205.

Khoo, H. E., Azlan, A., Tang, S. T., & Lim, S. M. (2017). Anthocyanidins and anthocyanins: colored pigments as food, pharmaceutical ingredients, and the potential health benefits. *Food and Nutrition Research*, 61(1), 1361779. https://doi.org/10.1080/16546628.2017.1361779.

Kim, H. K., Kim, J. N., Han, S. N., Nam, J. H., Na, H. N. & Ha, T. J. (2012a). Black soybean anthocyanins inhibit adipocyte differentiation in 3t3-L1 cells. *Nutrition Research,* 32(10), 770−777.

Kim, H. M., Kim, S. H. & Kang, B. S. (2018). Radioprotective effects of delphinidin on normal human lung cells against proton beam exposure. *Nutrition Research and Practice,* 12(1), 41−46.

Kim, H. S., Sul, D., Lim, J. Y., Lee, D., Joo, S. S., Hwang, K. W. & Park, S. Y. (2009). Delphinidin ameliorates beta-amyloid-induced neurotoxicity by inhibiting calcium influx and tau hyperphosphorylation. *Bioscience, Biotechnology and Biochemistry,* 73(7), 1685–1689.

Kim, S. M., Chung, M. J., Ha, T. J., Choi, H. N., Jang, S. J., Kim, S. O., Chun, M. H., Do, S. I., Choo, Y. K. & Park, Y. I. (2012b). Neuroprotective effects of black soybean anthocyanins via inactivation of Ask1-Jnk/P38 pathways and mobilization of cellular sialic acids. *Life Sciences,* 90(21–22), 874−882.

Knaup, B., Oehme, A., Valotis, A. & Schreier, P. (2009). Anthocyanins as lipoxygenase inhibitors. *Molecular Nutrition & Food Research,* 53, 617–624.

Kopustinskiene, D. M., Jakstas, V., Savickas, A. & Bernatoniene, J. (2020). Flavonoids as anticancer agents. Nutrients, 12(2), 457. https://doi.org/10.3390/nu12020457.

Korte, G., Dreiseitel, A., Schreier, P., Oehme, A., Locher, S., Hajak, G. & Sand P. G. (2009). An examination of anthocyanins' and anthocyanidins' affinity for cannabinoid receptors. *Journal of Medicinal Food*, 12, 1407–1410.

Krga, I., Monfoulet, L. E., Konic-Ristic, A., Mercier, S., Glibetic, M., Morand, C. & Milenkovic, D. (2016). Anthocyanins and their gut metabolites reduce the adhesion of monocyte to Tnfalpha-activated endothelial cells at physiologically relevant concentrations. *Archives of Biochemistry and Biophysics,* 599, 51−59.

Krga, I., Tamaian, R., Mercier, S., Boby, C., Monfoulet, L. E., Glibetic, M., Morand, C. & Milenkovic, D. (2018a). Anthocyanins and their gut metabolites attenuate monocyte adhesion and transendothelial migration through nutrigenomic mechanisms regulating endothelial cell permeability. *Free Radical Biology and Medicine,* 124, 364−379.

Krga, I., Vidovic, N., Milenkovic, D., Konic-Ristic, A., Stojanovic, F., Morand, C. & Glibetic, M. (2018b). Effects of anthocyanins and their gut metabolites on adenosine diphosphate-induced platelet activation and their aggregation with monocytes and neutrophils. *Archives of Biochemistry and Biophysics,* 645, 34−41.

Kuhn, I., Kellenberger, E., Said-Hassane, F., Villa, P., Rognan, D., Lobstein, A., Haiech, J., Hibert,M., Schuber,F. & Muller-Steffner, H. (2010). Identification by high-throughput screening of inhibitors of *Schistosoma mansoni* NAD(+) catabolizing enzyme. *Bioorganic and Medicinal Chemistry,* 18, 7900–7910.

Kumoro, Ac., Retnowati, D. S. & Sri Budiyati, C. (2010). Solubility of delphinidin in water and various organic solvents between 298.15 and 343.15 K. *Journal of Chemical and Engineering Data*, 55(7), 2603–2606.

Kwon, J. Y., Lee, K. W., Kim, J. E., Jung, S. K., Kang, N. J., Hwang, M. K., Heo, Y. S., Bode, A. M., Dong, Z. & Lee, H. J. (2009). Delphinidin suppresses ultraviolet B-induced cyclooxygenases-2 expression through inhibition of MAPKK4 and PI-3 kinase. *Carcinogenesis*, 30 (11), 1932–1940.

Lee, J. S., Kim, Y. R., Song, I. G., Ha, S., Kim, Y. E., Baek, N. & Hong, E. K. (2015). Cyanidin-3-glucoside isolated from mulberry fruit protects pancreatic β-cells against oxidative stress-induced apoptosis. *International Journal of Molecular Medicine*, 35, 405–412.

Lee, S. G., Kim, B., Yang, Y., Pham, T. X., Park, Y. K., Manatou, J., Koo, S. I., Chun, O. K. & Lee, J. Y. (2014). Berry anthocyanins suppress the expression and secretion of proinflammatory mediators in macrophages by inhibiting nuclear translocation of NF-κB independent of NRF2-mediated mechanism. *Journal of Nutritional Biochemistry*, 25(4), 404–411.

Legeay, S., Tran, K., Abatuci, Y., Justiniano, H., Lugnier, C., Duval, O., Helesbeux, J. J. & Faure, S. (2019). Design, synthesis, pharmacological evaluation and vascular effects of delphinidin analogues. *Current Pharmaceutical Design,* 24(46), 5580−5589.

León-González, A. J., Sharif, T., Kayali, A., Abbas, M., Dandache, I., Etienne-Selloum, N., Kevers, C., Pincemail, J., Auger, C., Chabert, P., Alhosin, M. & Schini-Kerth, V. B. (2015). Delphinidin-3-O-glucoside and delphinidin3-O-rutinoside mediate the redox-sensitive caspase 3-related pro-apoptotic effect of black currant juice on leukaemia jurkat cells. *Journal of Functional Foods*, 17, 847–856.

Lim, T. G., Jung, S. K., Kim, J. E., Kim, Y., Lee, H. J., Jang, T. S. & Lee, K. W. (2013). NADPH oxidase is a novel target of delphinidin for the inhibition of uvb-induced mmp-1 expression in human dermal fibroblasts. *Experimental Dermatology,* 22(6), 428−430.

Lin, B. W., Gong, C. C., Song, H. F. & Cui, Y. Y. (2017). Effects of anthocyanins on the prevention and treatment of cancer. *British journal of pharmacology*, 174(11), 1226–1243.

Lin, Y. C., Tsai, P. F. & Wu, J. S. (2014). Protective effect of anthocyanidins against sodium dithionite-induced hypoxia injury in c6 glial cells. *Journal of Agricultural and Food Chemistry,* 62(24), 5603−5608.

Liobikas, J., Skemiene, K., Trumbeckaite, S. & Borutaite, V. (2016). Anthocyanins in cardioprotection: A path through mitochondria. *Pharmacological Research,* 113(B), 808−815.

Long, L. H., Hoi, A. & Halliwell, B. (2010). Instability of, and generation of hydrogen peroxide by, phenolic compounds in cell culture media. *Archives of Biochemistry and Biophysics*, 501(1), 162–169.

Lugrin, J., Rosenblatt-Velin, N., Parapanov, R. & Liaudet, L. (2014). The role of oxidative stress during inflammatory processes. *Biological Chemistry* 395(2), 203−230.

Luna-Vital, D. A. & Gonzalez de Mejia, E. (2018). Anthocyanins from purple corn activate free fatty acid-receptor 1 and glucokinase enhancing in vitro insulin secretion and hepatic glucose uptake. *PLoS One*, 13(7), e0200449. https://doi.org/10.1371/journal.pone.0200449.

Matsunaga, N., Imai, S., Inokuchi, Y., Shimazawa, M., Yokota, S., Araki, Y. & Hara, H. (2009). Bilberry and its main constituents have neuroprotective effects against retinal neuronal damage in vitro and in vivo. *Molecular Nutrition and Food Research*, 53, 869–877.

Matsunaga, N., Tsuruma, K., Shimazawa, M., Yokota, S. & Hara, H. (2010). Inhibitory actions of bilberry anthocyanidins on angiogenesis. *Phytotherapy Research*, 1, S42–S47.

McCallum, J. L., Yang, R., Young, J. C., Strommer, J. N. & Tsao, R. (2007). Improved high performance liquid chromatographic separation of anthocyanin compounds from grapes using a novel mixed-mode ion-exchange reversed-phase column. *Journal of Chromatography A*, 1148(1), 38–45.

Mojica, L., Berhow, M. & Gonzalez de Mejia, E. (2017). Black bean anthocyanin-rich extracts as food colorants: Physicochemical stability and antidiabetes potential. *Food Chemistry*, 229, 628–639.

Moriwaki, S., Suzuki, K., Muramatsu, M., Nomura, A., Inoue, F., Into, T., Yoshiko, Y. & Niida, S. (2014). Delphinidin, one of the major anthocyanidins, prevents bone loss through the inhibition of excessive osteoclastogenesis in osteoporosis model mice. *PLoS One*, 9(5), e97177. https://doi.org/10.1371/journal.pone.0097177

Murata, M., Kosaka, R., Kurihara, K., Yamashita, S. & Tachibana, H. (2016). Delphinidin prevents disuse muscle atrophy and reduces stress related gene expression. *Bioscience, Biotechnology, and Biochemistry,* 80(8), 1636−1640.

Nam, D. C., Hah, Y. S., Nam, J. B., Kim, R. J. & Park, H. B. (2016). Cytoprotective mechanism of cyanidin and delphinidin against oxidative stress-induced tenofibroblast death. *Biomolecules and Therapeutics,* 24(4), 426−432.

Nanashima, N., Horie, K. & Maeda, H. (2018). Phytoestrogenic activity of blackcurrant anthocyanins is partially mediated through estrogen receptor beta. *Molecules*, 23(1), 74. https://doi.org/10.3390/molecules23010074.

Nanashima, N., Horie, K., Tomisawa, T., Chiba, M., Nakano, M., Fujita, T., Maeda, H., Kitajima, M., Takamagi, S., Uchiyama, D., Watanabe, J., Nakamura, T. & Kato, Y. (2015). Phytoestrogenic activity of blackcurrant (*Ribes nigrum*) anthocyanins is mediated through estrogen receptor alpha. *Molecular Nutrition and Food Research*, 59(12), 2419−2431.

Naseri, R., Farzaei, F., Haratipour, P., Nabavi, S. F., Habtemariam, S., Farzaei, M. H., Khodarahmi, R., Tewari, D. & Momtaz, S. (2018). Anthocyanins in the management of metabolic syndrome: A pharmacological and biopharmaceutical review. *Frontiers in Pharmacology,* 9, 1310. https://doi.org/10.3389/fphar.2018.01310.

Ni, T., Yang, W. & Xing, Y. (2019). Protective effects of delphinidin against H_2O_2-induced oxidative injuries in human retinal pigment epithelial cells. *Bioscience Reports,* 39(8), BSR20190689. https://doi.org/10.1042/BSR20190689.

Nyanhanda, T., Gould, E. M., McGhie, T., Shaw, O. M., Harper, J. L. & Hurst, R. D. (2014). Blackcurrant cultivar polyphenolic extracts suppress ccl26 secretion from alveolar epithelial cells. *Food Function,* 5(4), 671−677.

Ogawa, K., Oyagi, A., Tanaka, J., Kobayashi, S. & Hara, H. (2011). The protective effect and action mechanism of *Vaccinium myrtillus* L. on gastric ulcer in mice. *Phytotherapy Research*, 25, 1160–1165

Ostberg-Potthoff, J. J., Berger, K., Richling, E. & Winterhalter, P. (2019). Activity-guided fractionation of red fruit extracts for the identification of compounds influencing glucose metabolism. *Nutrients*, 11(5), 1166. https://doi.org/10.3390/nu11051166.

Ozbay, T. & Nahta, R. (2011). Delphinidin inhibits HER2 and Erk1/2 signaling and suppresses growth of HER2-overexpressing and triple negative breast cancer cell lines. *Breast Cancer*; 5, 143–154.

Paixão, J., Dinis, T. C. & Almeida, L. M. (2011). Dietary anthocyanins protect endothelial cells against peroxynitrite-induced mitochondrial apoptosis pathway and Bax nuclear translocation: An in vitro approach. *Apoptosis,* 16, 976-–989.

Pal, H. C., Chamcheu, J. C., Adhami, V. M., Wood, G. S., Elmets, C. A., Mukhtar, H. & Afaq, F. (2015). Topical application of delphinidin reduces psoriasiform lesions in the flaky skin mouse model by inducing epidermal differentiation and inhibiting inflammation.*British Journal of Dermatology*, 172(2), 354–364.

Parichatikanond, W., Pinthong, D. & Mangmool, S. (2012). Blockade of the renin-angiotensin system with delphinidin, cyanin, and quercetin. *Planta Medica*, 78(15), 1626−1632.

Park, M., Sharma, A. & Lee, H. J. (2019). Anti-adipogenic effects of delphinidin-3-o-beta-glucoside in 3t3-l1 preadipocytes and primary white adipocytes. *Molecules*, 24(10), 1848. https://dx.doi.org/10.3390/molecules24101848.

Parra-Vargas, M., Sandoval-Rodriguez, A., Rodriguez-Echevarria, R., Dominguez-Rosales, J. A., Santos-Garcia, A. & Armendariz-Borunda, J. (2018). Delphinidin ameliorates hepatic triglyceride accumulation in human Hepg2 Cells, but not in diet-induced obese mice. *Nutrients*, 10(8), 1060. https://dx.doi.org/10.3390/nu10081060.

Patel, K., Jain, A., Patel, D. K. (2013).Medicinal significance, pharmacological activities, and analytical aspects of anthocyanidins 'delphinidin': A concise report.*Journal of Acute Disease,* 2(3), 169–178.

Pei, L., Wan, T., Wang, S., Ye, M., Qiu, Y., Jiang, R., Pang, N., Huang, Y., Zhou, Y., Jiang, X., Ling, W., Zhang, Z. & Yang, L. (2018). Cyanidin-3-O-beta-glucoside regulates the activation and the secretion of adipokines from brown adipose tissue and alleviates diet induced fatty liver. *Biomedicine and Pharmacotherapy,* 105, 625−632.

Quintos, L., Lee, I. A., Kim, H. J., Lim, J. S., Park, J., Sung, M. K., Seo, Y. R. & Kim, J. S. (2010). Significance of p27 as potential biomarker for intracellular oxidative status. *Nutrition Research and Practice,* 4, 351-355.

Rahman, N., Jeon, M. & Kim, Y. S. (2016). Delphinidin: A major anthocyanin, inhibits 3t3-l1 pre-adipocyte differentiation through activation of wnt/beta-catenin signaling. *Biofactors*, 42(1), 49−59.

Rat'kin, A. V. & Tarasov, V. A. (2010). Genetic control of biosynthesis of anthocyans in sweet pea (*Lathyrus odoratus* L.) flowers. *Genetika*, 46, 488–496.

Real Hernandez, L. M., Fan, J., Johnson, M. H. & Gonzalez de Mejia, E. (2015). Berry phenolic compounds increase expression of hepatocyte nuclear factor-1alpha (Hnf-1alpha) in caco-2 and normal colon cells due to high affinities with transcription and dimerization domains of Hnf-1alpha. *PLoS One*, 10 (9), e0138768. https://dx.doi.org/10.1371/journal.pone.0138768.

Ricciotti, E. & Fitz-Gerald, G. A. (2011). Prostaglandins and inflammation. *Arteriosclerosis, Thrombosis, and Vascular Biology,* 31(5), 986−1000.

Sakaguchi, K., Kitajima, J. & Iwashina, T. (2013). Acylated delphinidin glycosides from violet and violet-blue flowers of *Clematis cultivars* and their coloration. *Natural Product Communications,* 8(11), 1563–1566.

Sakaki, J. R., Melough, M. M. & Chun, O. K. (2020).Anthocyanins and anthocyanin-rich food as antioxidants in bone pathology. *In: Pathology, Academic Press,* pp. 145–158.

Santos, J., La, V. D., Bergeron, C. & Grenier, D. (2011). Inhibition of host- and bacteria-derived proteinases by natural anthocyanins. *Journal of Periodontal Research*, 46(5), 550–557.

Satué-Gracia, M. T., Heinonen, M. & Frankel, E. N. (1997). Anthocyanins as antioxidants on human low-density lipoprotein and lecithin-liposome systems. *Journal of Agricultural and Food Chemistry,* 45, 3362–3367.

Saulite, L., Jekabsons, K., Klavins, M., Muceniece, R. & Riekstina U. (2019). Effects of malvidin, cyanidin and delphinidin on human adipose mesenchymal stem cell differentiation into adipocytes, chondrocytes and osteocytes. *Phytomedicine,* 53, 86–95.

Scarabelli, T. M., Mariotto, S., Abdel-Azeim, S., Shoji, K., Darra, E., Stephanou, A., Chen-Scarabelli, C., Marechal, J. D., Knight, R., Ciampa, A., Saravolatz, L., de Prati, A. C., Yuan, Z., Cavalieri, E., Menegazzi, M., Latchman, D., Pizza, C., Perahia, D. & Suzuki, H. (2009). Targeting stat1 by myricetin and delphinidin provides efficient protection of the heart from ischemia/reperfusion-induced injury. *FEBS Letters*, 583(3), 531–541.

Seeram, N. P., Zhang, Y. & Nair, M. G. (2003). Inhibition of proliferation of human cancer cells and cyclooxygenase enzymes by anthocyanidins and catechins. *Nutrition and Cancer,* 46(1), 101–106.

Sehrawat, N., Yadav, M., Kumar, S. Upadhyay, S. K., Singh, M. & Sharma, A. K. (2020). Review on health promoting biological activities of mungbean: A potent functional food of medicinal importance. *Plant Archives*, 20(2), 2969–2975.

Seo, B. N., Ryu, J. M., Yun, S. P., Jeon, J. H., Park, S. S., Oh, K. B., Park, J. K. & Han, H. J. (2013). Delphinidin prevents hypoxia-induced mouse embryonic stem cell apoptosis through reduction of intracellular reactive oxygen species-mediated activation of Jnk and Nf-Kappab, and Akt Inhibition. *Apoptosis*, 18(7), 811–824.

Seong, A. R., Yoo, J. Y., Choi, K., Lee, M. H., Lee, Y. H., Lee, J., Jun, W., Kim, S. & Yoon, H. G. (2011). Delphinidin, a Specific inhibitor of histone acetyltransferase, suppresses inflammatory signaling via prevention of Nf-Kappab acetylation in fibroblast-like synoviocyte Mh7a cells. *Biochemical and Biophysical Research Communication,* 410(3), 581–586.

Silvan, J. M., Reguero, M. & de Pascual-Teresa, S. A (2016). Protective effect of anthocyanins and xanthophylls on uvb-induced damage in retinal pigment epithelial cells. *Food and Function*, 7(2), 1067–1076.

Singh, C. K., Siddiqui, I. A., El-Abd, S., Mukhtar, H. & Ahmad, N. (2016). Combination chemoprevention with grape antioxidants. *Molecular Nutrition and Food Research,* 60, 1406–1415.

Singletary, K. W., Jung, K. J. & Giusti, M. (2007). Anthocyanin-rich grape extract blocks breast cell DNA damage.*Journal of Medicinal Food*, 10(2), 244–251.

Skemiene, K., Liobikas, J. & Borutaite, V. (2015). Anthocyanins as substrates for mitochondrial complex i - Protective effect against heart ischemic injury. *FEBS Journal,* 282(5), 963–971.

Sogo, T., Terahara, N., Hisanaga, A., Kumamoto, T., Yamashiro, T., Wu, S., Sakao, K. & Hou, D. X. (2015). Anti-inflammatory activity and molecular mechanism of delphinidin 3-sambubioside, a hibiscus anthocyanin. *BioFactors*, 41(1), 58–65.

Song, S. E., Jo, H. J., Kim, Y. W., Cho, Y. J., Kim, J. R. & Park, S. Y. (2016). Delphinidin prevents high glucose-induced cell proliferation and collagen synthesis by inhibition of Nox-1 and mitochondrial superoxide in mesangial cells. *Journal of Pharmacological Sciences,* 130(4), 235–243.

Sorrenti, V., Vanella, L., Acquaviva, R., Cardile, V., Giofrè, S. & Di Giacomo, C. (2015). Cyanidin induces apoptosis and differentiation in prostate cancer cells. *International Journal of Oncology,* 47, 1303–1310.

Srivastava, C., Akoh, C., Fischer, J. & Krewer, G. (2007). Effect of anthocyanin fractions from selected cultivars of Georgia-grown blueberries on apoptosis and phase II enzymes. *Journal of Agricultural and Food Chemistry,* 55, 3180–3185.

Stevens, M., Neal, C. R., Craciun, E. C., Dronca, M., Harper, S. J. & Oltean, S. (2019). The natural drug Diavit is protective in a type ii mouse model of diabetic nephropathy. *PLoS One,* 14(3), e0212910. https://doi.org/10.1371/journal.pone.0212910.

Sun, C. D., Zhang, B., Zhang, J. K., Xu, C. J., Wu, Y. L., Li, X. & Chen, K. S. (2012a). Cyanidin-3-glucoside-rich extract from Chinese bayberry fruit protects pancreatic β cells and ameliorates hyperglycemia in streptozotocin induced diabetic mice. *Journal of Medicinal Food,* 15, 288–298.

Sun, D., Huang, J., Zhang, Z., Gao, H., Li, J., Shen, M., Cao, F. & Wang, H., (2012b). Luteolin limits infarct size and improves cardiac function after myocardium ischemia/reperfusion injury in diabetic rats. *PLoS One,* 7(3), e33491. https://doi.org/10.1371/journal.pone.0033491.

Takasawa, R., Saeki, K., Tao, A., Yoshimori, A., Uchiro, H., Fujiwara, M. & Tanuma, S. I. (2010). Delphinidin, a dietary anthocyanidin in berry fruits, inhibits human glyoxalase I. *Bioorganic and Medicinal Chemistry*, 18, 7029–7033.

Takayanagi, H., Kim, S., Koga, T., Nishina, H., Isshiki, M., Yoshida, H., Saiura, A., Isobe, M., Yokochi, T., Inoue, J., Wanger, E. F., Mak, T. W., Kodama, T. & Taniguchi, T. (2002). Induction and activation of the transcription factor NFATc1 (NFAT2) integrate RANKL signaling in terminal differentiation of osteoclasts. *Developmental Cell,* 3, 889–901.

Tamura, S., Yoshihira, K., Fujiwara, K. & Murakami, N. (2010). New inhibitors for expression of Ige receptor on human mast cell. *Bioorganic and Medicinal Chemistry Letters,* 20(7), 2299–2302.

Tanaka, Y., Sasaki, N. & Ohmiya, A. (2008). Biosynthesis of plant pigments: Anthocyanins, betalains and carotenoids. *Plant Journal for Cell and Molecular Biology*, 54(4), 733–749.

Tani, T., Nishikawa, S., Kato, M. & Tsuda, T. (2017). Delphinidin 3- rutinoside-rich blackcurrant extract ameliorates glucose tolerance by increasing the release of glucagon-like peptide-1 secretion. *Food Scienceand Nutrition,* 5(4), 929–933.

Teller, N., Thiele, W., Boettler, U., Sleeman, J. & Marko, D. (2009). Delphinidin inhibits a broad spectrum of receptor tyrosine kinases of the ErbB and VEGFR family. *Molecular Nutrition and Food Research*, 53, 1075–1083.

Thilavech, T., Ngamukote, S., Belobrajdic, D., Abeywardena, M. & Adisakwattana, S. (2016). Cyanidin-3-rutinoside attenuates methylglyoxal-induced protein glycation and DNA damage via carbonyl trapping ability and scavenging reactive oxygen species. *BMC Complementaryand Alternative Medicine,* 16, 138. https://doi.org/10.1186/s12906-016-1133-x.

Tsutsumi, A., Horikoshi, Y., Fushimi, T., Saito, A., Koizumi, R., Fujii, Y., Hu, Q. Q., Hirota, Y., Aizawa, K. & Osakabe, N. (2019). Acylated anthocyanins derived from purple carrot (*Daucus carota* L.) induce elevation of blood flow in rat cremaster arteriole. *Food and Function*, 10(3), 1726−1735.

Tuli, H. S., Mittal, S., Aggarwal, D., Parashar, G., Parashar, N. C., Upadhyay, S. K., Barwal, T.S., Jain, A., Kaur, G., Salva, R., Sak, K., Kumar, M., Varol., M., Iqubal, A. & Sharma, A.K. (2021). Path of silibinin from diet to medicine: A dietary polyphenolic flavonoid having potential anti-cancer therapeutic significance. *Seminars in Cancer Biology*, 73, 196–218.

Venancio, V. P., Cipriano, P. A., Kim, H., Antunes, L. M., Talcott, S. T. & Mertens-Talcott, S. U. (2017). Cocoplum (*Chrysobalanus icaco* L.) anthocyanins exert anti-inflammatory activity in human colon cancer and non-malignant colon cells. *Food and Function*, 8(1), 307−314.

Wada, T., Nakashima, T., Hiroshi, N. & Penninger, J. M. (2006) RANKL-RANK signaling in osteoclastogenesis and bone disease. *Trends in Molecular Medicine,* 12, 17–25.

Wallace, T. C. (2011). Anthocyanins in cardiovascular disease. *Advances in Nutrition*, 2(1), 1–7.

Wang, C. H., Zhu, L. L., Ju, K. F., Liu, J. L. & Li, K. P. (2017). Anti-inflammatory effect of delphinidin on intramedullary spinal pressure in a spinal cord injury rat model. *Experimental and TherapeuticMedicine,* 14(6), 5583−5588.

Wang, H., Cao, G. & Prior, R.L. (1997). Oxygen radical absorbing capacity of anthocyanins. *Journal of Agricultural and Food Chemistry,* 45, 304–309.

Wang, Y., Zhao, L., Wang, D., Huo, Y. & Ji, B. (2016). Anthocyanin-rich extracts from blackberry, wild blueberry, strawberry, and chokeberry: Antioxidant activity and inhibitory effect on oleic acid-induced hepatic steatosis *in vitro*. *Journal of the Science of Food and Agriculture*,96(7), 2494–2503.

Xie, X., Zhao, R. & Shen, G. X. (2012). Influence of delphinidin-3- glucoside on oxidized low-density lipoprotein-induced oxidative stress and apoptosis in cultured endothelial cells. *Journal of Agricultural and Food Chemistry,* 60(7), 1850−1856.

Yadav, M., Sehrawat, N., Singh, M., Upadhyay, S. K., Aggarwal, D. & Sharma, A. K. (2020). Cardioprotective and hepatoprotective potential of citrus flavonoid naringin: Current status and future perspectives for health benefits. *Asian Journal of Biological and Life Sciences,* 9(1), 1–5.

Yadav, M., Sehrawat, N., Upadhyay, S. K. & Kumar, S. (2021a). Emerging renoprotective role of citrus flavonoid naringin: Current pharmaceutical status and future perspectives. *Current Pharmacology Reports*, 7(3), 96–101.

Yadav, M., Upadhyay, S. K., Singh, R., Sharma, A. K., Kumar, S., Rani, K. & Sehrawat, N. (2021b). Effects of salinity stress on seedling growth of local varieties of faba bean (*Vicia faba* L.) from India. *Plant Archives*, 21(1), 253–259.

Yamamoto, E., Nishimura, N., Okada, K., Sekido, C., Yamamichi, S. & Hasumi, K. (2011). Inhibitors of autoactivation of plasma hyaluronanbinding protein (factor VII activating protease). *Biological and Pharmaceutical Bulletin,* 34, 462-470.

Yang, J. B., Kang, Y. M., Zhang, C., Yu, X. J. & Chen, W. S. (2019). Infusion of melatonin into the paraventricular nucleus ameliorates myocardial ischemia-reperfusion injury by regulating oxidative stress and inflammatory cytokines. *Journal of Cardiovascular Pharmacology,* 74(4), 336–347.

Yang, Y., Heo, P., Kong, B., Park, J. B., Jung, Y. H., Shin, J., Jeong, C. & Kweon, D. H. (2015). Dynamic light scattering analysis of snare driven membrane fusion and the effects of snare-binding flavonoids. *Biochemical and Biophysical Research Communications,* 465(4), 864−870.

Yang, Y., Shin, J. Y., Oh, J. M., Jung, C. H., Hwang, Y., Kim, S., Kim, J. S., Yoon, K. J., Ryu, J. Y., Shin, J., Hwang, J. S., Yoon, T. Y., Shin, Y. K. & Kweon, D. H. (2010). Dissection of snare-driven membrane fusion and neuroexocytosis by wedging small hydrophobic molecules into the snare zipper. *Proceedings of the National Academy of Sciences of the United States of America,* 107 (51), 22145−22150.

Yao, Y., Wang, Y., Zhang, Y. & Liu, C. (2017). Klotho ameliorates oxidized low density lipoprotein (ox-LDL)-induced oxidative stress via regulating LOX-1 and PI3K/Akt/eNOS pathways.*Lipids in Health and Disease*, 16(1), 77. https://doi.org/10.1186/s12944-017-0447-0.

Yi, L., Chen, C. Y., Jin, X., Zhang, T., Zhou, Y., Zhang, Q. Y., Zhu, J. D. & Mi, M. T. (2012). Differential suppression of intracellular reactive oxygen species-mediated signalingpathway in vascular endothelial cells by several subclasses of flavonoids. *Biochimie*, 94(9), 2035−2044.

Yun, J. M., Afaq, F., Khan, N. & Mukhtar, H. (2009). Delphinidin, an anthocyanidin in pigmented fruits and vegetables, induces apoptosis and cell cycle arrest in human colon cancer HCT116 cells. *Molecular Carcinogenesis*, 48, 260–270.

Zamora-Ros, R., Knaze, V., Lujan-Barroso, L., Slimani, N., Romieu, I., Touillaud, M., Kaaks, R., Teucher, B., Mattiello, A., Grioni, S., Crowe, F., Boeing, H., Forster, J., Quiros, J. R., Molina, E., Huerta, J. M., Engeset, D., Skeie, G., Trichopoulou, A., Dilis, V., Tsiotas, K., Peeters, P. H., Khaw, K. T., Wareham, N., Bueno-DeMesquita, B., Ocke, M. C., Olsen, A., Tjonneland, A., Tumino, R., Johansson, G., Johansson, I., Ardanaz, E., Sacerdote, C., Sonestedt, E., Ericson, U., Clavel-Chapelon, F., Boutron-Ruault, M. C., Fagherazzi, G., Salvini, S., Amiano, P., Riboli, E. & Gonzalez, C. A. (2011). Estimation of the intake of anthocyanidins and their food sources in the european prospective investigation into cancer and nutrition (epic) study. *British Journal of Nutrition,* 106(7), 1090−1099.

Zhao, C. L., Yu, Y. Q., Chen, Z. J., Wen, G. S., Wei, F. G., Zheng, Q., Wang, C. D. & Xiao, X. L. (2017). Stability-increasing effects of anthocyanin glycosyl acylation. *Food Chemistry*, 214, 119−128.

Zhong, S., Sandhu, A., Edirisinghe, I. & Burton-Freeman, B. (2017). Characterization of wild blueberry polyphenols bioavailability and kinetic profile in plasma over 24-h period in human subjects. *Molecular Nutrition and Food Research*, 61(12), 1700405. https://doi.org/10.1002/mnfr.201700405.

About the Editors

Dr. Nirmala Sehrawat

She is serving as Assistant Professor at Department of Biotechnology, Maharishi Markandeshwar (Deemed to be University), Mullana-Ambala, India. She has more than 10 years of teaching and research experience. She accomplished her B.Sc. and M.Sc. in Biotechnology from Maharshi Dayanand University, Rohtak, India. She obtained her Ph.D. degree from M.D. University, Rohtak, India. She has qualified the CSIR-NET and ICAR-NET. She has published more than 70 research and review articles in various reputed journals. She is currently working on food grain legumes and their industrial aspects including mungbean as functional food. She is active member of several professional and scientific societies. She is also serving as editorial board member for various national and international journals.

Dr. Diwakar Aggarwal

He is presently working as Assistant Professor in Department of Biotechnology, Maharishi Markandeshwar (Deemed to be University),

Mullana-133207, Ambala, Haryana (India). He has 18 years of teaching and research experience. He has done B.Sc. with specialization in Industrial Microbiology from Kurukshetra University, Kurukshetra and did his M.Sc. and Ph.D. in Biotechnology from Thapar University, Patiala (Punjab), India. Dr. Aggarwal has published more than 50 papers including book chapters in journals/books of international repute and has more than 700 citations with an h-index of 14.

Dr. Sushil Kumar Upadhyay

Assistant Professor in Biotechnology at Maharishi Markandeshwar (Deemed to be University), Mullana-Ambala (HR). Dr. Upadhyay is Doctorate from University of Allahabad (*Oxford of East*), has 14+ years of rich teaching-cum-research experience in Animal & Veterinary Science, Parasitology, Phylogenetics, Nanobiotechnology & Microbiology. Dr. Upadhyay has received more the 2 dozen awards including certificate of excellence, Bharat Jyoti award, Bharat Shiksha Gaurav Puraskar, Prof. S. Yamuguti Parasites Taxonomy Promotion award, Prof. S.P. Gupta Gold medal, Prof. S.L. Mishra Gold medal, Best Teacher award, Paryavaran Shri Samman, Paryavaran Ratna Puraskar, Vigyan Ratna Puraskar and felicitated many Fellow awards as FISEC, FSESc, FSSc, FSLSc, FISCA, FZSI, FAGEM, FHSI, FMERC, FBS, FSEZR, FIAZ, FGESA, FI2OR, FGT, FSAN, FSASS, FAELS, FSOE, FLS (London) in yesteryears. Dr. Upadhyay has served as Special Officer WCCB GoI, Global Goodwill Ambassador India (Humanitarian), Sustainable Cosmos Ambassador GTS in past years. His research skill is excellent; he has published 9 patents, 31 books and 135 research articles in high impact journals, attended 57 FDPs/OC/RC and presented papers at 185 conferences also. Dr. Upadhyay has left no stone unturned in the field of Mol-Taxonomy and Phylogeny for taxonomic validation of parasitic helminthes. His contribution for economic empowerment to youths at grassroots level for livelihood and sustainability by mission "*lab to land transfer of science and technology*" and vision "*earn as you learn*" is also remarkable and can't be ignored. Dr. Upadhyay is an ideal teacher and his hard work inspires students to achieve their goal be 'SMART' [S=Specific, M=Meaningful, A=Achievable, R=Relevant, T=Time-bound].

Dr. Mukesh Yadav

Assistant Professor at the Department of Biotechnology, Maharishi Markandeshwar (MM) (Deemed to be University), Mullana-Ambala, India. He accomplished his B.Sc. and M.Sc. in Biotechnology from Maharshi Dayanand University, Rohtak, India. He completed his Ph.D. degree in Microbial Biotechnology from Punjabi University, Patiala, India. He has been awarded with Indian fellowship programs (CSIR and DBT). His area of expertise is microbial biotechnology. He is extensively involved in research domain of fermentative production of microbial enzymes and microbial metabolites of industrial importance. He has published more than 70 research and review articles in various reputed national and international journals. He has published books on microbial enzymes for food industries, industrial biotechnology, and therapeutic potential of flavonoids and instrumentation. He has life membership of several professional scientific societies and contributes to the development of society.

Index

B

C

D

K

L

M

N

T

U

V

W

X

Y